Measuring the Economic Impacts of Transgenic Crops in Developing Agriculture during the First Decade

Approaches, Findings, and Future Directions

Food Policy Review 10

IFPRI Food Policy Reviews

The Food Policy Review series provides a forum for scholarly reviews of the literature on key topics relevant to food and agricultural policy. These works are substantive syntheses of research on major food policy issues and may include original research to supplement the discussion. They provide considered judgments on policy options in light of current knowledge of the processes underlying the reduction of poverty, hunger, and malnutrition in the world.

All manuscripts submitted for publication in the Food Policy Review series undergo extensive review under the auspices of IFPRI's Publications Review Committee. The review process includes selection by the Review Committee of external and internal reviewers chosen for their expertise in the manuscript's subject matter and methodology. The author responds to the reviewer's comments, and, if approved by the Review Committee and IFPRI's director general, the revised manuscript becomes part of the IFPRI Food Policy Review series.

About IFPRI

The International Food Policy Research Institute (IFPRI) has a vision of a world in which every person has secure access to sufficient and safe food to sustain a healthy and productive life and in which food-related policy decisions are made transparently and include the participation of consumers and producers. In pursuit of these goals, IFPRI identifies and analyzes international, national, and local policies in support of improved food security and nutrition; strengthens the capabilities of people and institutions in developing countries that conduct food policy research; and communicates research results to those in a position to apply or use them.

Established in 1975, IFPRI is one of 15 agricultural research centers that receive their principal funding from governments, private foundations, and international and regional organizations,

Measuring the Economic Impacts of Transgenic Crops in Developing Agriculture during the First Decade

Approaches, Findings, and Future Directions

Melinda Smale, Patricia Zambrano, Guillaume Gruère,
José Falck-Zepeda, Ira Matuschke, Daniela Horna,
Latha Nagarajan, Indira Yerramareddy, and Hannah Jones

Food Policy Review 10

INTERNATIONAL FOOD POLICY RESEARCH INSTITUTE
sustainable solutions for ending hunger and poverty
IFPRI®
Supported by the CGIAR

International Food Policy Research Institute
2033 K Street, NW
Washington, D.C. 20006-1002, U.S.A.
Telephone +1-202-862-5600
www.ifpri.org

DOI: 10.2499/0896295117FPRev10

Library of Congress Cataloging-in-Publication Data

Measuring the economic impacts of transgenic crops in developing agriculture during the first decade : approaches, findings, and future directions / Melinda Smale ... [et al.].
p. cm. — (Food policy review ; 10)
Includes bibliographical references.
ISBN 978-0-89629-511-7 (alk. paper)
1. Transgenic plants—Research—Developing countries. 2. Transgenic plants—Economic aspects—Developing countries. I. Smale, Melinda. II. Series: Food policy review ; 10.
SB123.57.M42 2009
338.1′6—dc22 2009003568

Contents

Tables

Foreword

Biotechnology in agriculture has generated a great deal of controversy in recent years. Of the many scientific advances that have occurred in plant breeding since Gregor Mendel conducted his experiments about 150 years ago, crops with genetic modifications seem to have been accorded a unique status. The use of crops that are modified by the transfer of genes across species has provoked concerns that continue to be echoed in the media and the academic press and have reached into the fields and lives of farmers in both rich and poor countries. An issue that remains unresolved is that what consumers and producers in rich countries may want is not necessarily what producers and consumers in poor countries may need (and want); hence, the preferences of the rich countries—transformed into science and development policies—may hinder the poor's access to needed technologies.

This review of scholarly literature explores a key concern of IFPRI's: whether biotech crops can benefit poor farmers. The authors examine the issue by emphasizing the methods applied to empirical data from developing countries, because these methods influence the nature of economists' findings and how they interpret them. The authors consider the economic impacts of biotech crops not only on farmers, but also on consumers, the agricultural sector as a whole, and international trade. They have also compiled a web-bibliography, bEcon, which is available to researchers, particularly those in developing countries, as a tool to further their own understanding of the evidence.

The authors conclude that biotech crops have promise for poor farmers. Further in-depth investigation is required. Bt cotton is by far the most studied biotech crop, but analysis of the economic impacts of other crops has only begun. Impacts on poverty, inequality, health, and the environment need more rigorous exploration.

Particular aspects of biotech crops—such as the institutional organization of their supply, the way that knowledge and transgenic seed are diffused in communities, and the costs and benefits of biosafety regulations—warrant in-depth investigation. So far, the published economics research that has applied a clearly identified method to empirical data collected in the fields of farmers in developing countries is limited. One reason is that few biotech crops have been introduced in developing-country agriculture, partly due to slow or hindered bio-policies and regulatory frameworks. Development cooperation organizations have not sufficiently invested in these; the above-mentioned preferences of some rich countries come into play here again. IFPRI is assisting developing countries to develop and implement such appropriate regulatory frameworks.

I trust that this report and the related web-bibliography will assist developing-country researchers in establishing their own evidence base and will help in their endeavors and encourage them to address the important questions that remain to be answered. Agricultural productivity and environmental challenges—including climate change—and growing long-term food needs will require access to and utilization of advanced biotechnology in developing-country agriculture.

Joachim von Braun
Director General, IFPRI

Acknowledgments

This study was made possible through the financial support of the International Development Research Centre, Canada; the European Union; and the World Bank. A collaborative project with Oxfam-America, led by Kimberly Pfeifer, contributed substantively to the research. The opinions expressed are those of the authors, and the authors alone are responsible for any errors of finding or interpretation. Patricia Zambrano, Indira Yerramareddy, and Hannah Jones conducted the systematic search on which this review is based. Mélodie Cartel conducted the initial search of the literature published in French. The results of the search, including articles referenced in this study, can be found in bEcon (http://www.ifpri.org/pubs/becon/becon.asp). bEcon is an interactive, public web-based bibliography maintained by IFPRI.

Summary

In the debate over biotech crops, differentiating fact from fiction is not easy. The debate has been confused by the influence of rigid, absolutist views (both supportive of and opposed to biotech crops) about the role of science in society, combined with a general ignorance of science. On the one hand, we hear that transgenic methods offer the chance to overcome some of the most intractable problems faced by poor farmers in harsh growing environments, such as drought. On the other, we hear concerns about the risks to human health and threats to biodiversity. Concentration of advanced scientific knowledge and market share in life-science corporations has provoked suspicion that poor farmers may have no say in the matter and cannot afford to purchase biotech seed anyway. Profound ethical issues, and skepticism about the benefits of transgenic crops, have led consumers and advocacy groups to vigorously resist adoption of biotech crops.

While we cannot claim objectivity on a topic fraught with such strongly held views, this systematic review represents our best effort to disentangle from the controversy some facts about the economic impacts of biotech crops on farmers in developing economies during the first decade of their use, 1997–2007. Although a number of reviews were published during that period, they focused only on findings or on one field of economic inquiry. Recognizing that methods applied by economists influence the nature of their findings and how they interpret them, this review focuses on methods. Analysis is presented in four chapters, each corresponding to a progressively larger actor in the agricultural economy: farmers, consumers, the crop industry or sector as a whole, and the nations and sectors engaged in international agricultural trade.

In the text, we have generally used the adjective "biotech" to describe a crop or product rather than GM (genetically modified) or transgenic (carrying a transgene inserted from another species). This term, which refers only to a type of technology, seems less provocative than the other terms after a decade of noisy debate.

An initial chapter lays out the criteria that were employed to delineate the boundaries of the literature search and presents tabular summaries of the search results. French-, Spanish-, and English-language literature was included. Only peer-reviewed literature with a stated method applied to empirical data from developing economies was examined. Thought pieces, essays, and polemics, as well as theoretical approaches, were omitted. Out of a vast literature about the economics of biotechnology, the number of articles and papers produced from 1997 to 2007 that met these criteria totals only 137. In other words, relatively little evidence is actually available on which to base generalizations. This is especially true when considering that the vast majority of the farm impact studies treat only one crop-trait combination (Bt cotton) in one of three countries (China, India, and South Africa); that most consumer studies have been conducted in China; and that many sector and trade models are by definition ex ante (before the fact) rather than ex post (after the fact). The number of authors is also far fewer than the number of studies, so that the range of voices in the published literature—with the notable exception of Indian studies—is still fairly narrow. In addition, even though all work examined has been subjected to some peer review, the range in quality is great.

A review of the methods employed to assess farm impacts in developing economies during the first decade confirms that evidence is not synonymous with fact. Two principal approaches have been used. The first, partial budgeting, indicates the marginal changes in variable costs and benefits per hectare that result from adopting biotech crops. In this approach, Bt cotton and Bt maize are hypothesized to increase yields, reduce insect damage, and reduce the use and costs of insecticides, including labor needs; herbicide-tolerant soybeans are expected to reduce the use of tillage and toxic herbicides. The second approach, which involves econometric models of varying specifications, tests these same hypotheses more rigorously.

The principal limits of partial budgeting are well known. Many studies report only gross margins, ignoring land and household labor. The treatment of risk through stochastic budgeting has been rare, despite its importance. Sample sizes have often been exceedingly small, and the debate over biotech crops has extended to the field—making it difficult to follow "best practices" with respect to sampling. Generally, estimates of quantities of labor, insecticides, and herbicides have been based not on measurement but on farmer recall, which is highly prone to error, especially for these inputs. More troubling is that sample selection bias is evident in a number of the early studies and has been addressed with statistical rigor only by a few more recent econometric approaches. Application of damage abatement models constituted a major advance over yield response models for Bt crops. Identifying the appropriate counterfactual to compare with the performance of the Bt variety or hybrid has posed challenges, since farmers do not generally also grow isogenic lines. In India, for example, the strong yield effect of host germplasm, aside from the effect of the transgene, was confirmed by econometric analyses of farm data. Another challenge that has not often been adequately addressed is the endogeneity of a farmer's decision to grow a Bt crop and apply insecticides. The same factors that affect yields also affect these decisions, which can lead to bias in estimated coefficients.

Rising adoption rates of Bt cotton and Bt maize among smallholders in a number of developing economies are a fact, as are the high adoption rates of herbicide-tolerant soybeans across Latin America, on farms that are much larger in scale. The evidence is generally conclusive that Bt cotton reduces pest damage and insecticide use and increases yields, but the magnitude of profits is highly variable, particularly in India. A notable exception is a set of studies by a researcher who used leaf tissue samples, stochastic analysis of partial budgets, and an advanced modeling approach to measure damage abatement.

Given the relatively short period of time since biotech crops were first adopted, the focus of field research has been to establish whether or not it pays for a farmer to purchase biotech seed in various farming contexts and to document farmers' use of this technology. Now, in the second decade of biotech use, the substance of empirical research needs to shift. Field researchers must use the most advanced methods at their disposal to address those impacts that most interest the public and policymakers: impacts on poverty and inequality in farming communities, effects on human health and the environment, implications for farmer knowledge systems and extension, and implications for female farmers, as compared with male farmers. Biotech crops have particular implications for the transfer of knowledge and the organization of seed supply and related information, as well as the empowerment of farmers and farming communities. These are some of the issues the public demands to hear about.

Turning to consumers, the literature remains circumscribed by the fact that few biotech food products have been released in developing economies. China is by far the most heavily studied developing economy in this literature, which includes three broad categories of research. The first consists of studies that report the findings of attitude surveys. These are generally descriptive in nature. In the second category, researchers exploit choice theory to model the willingness of consumers to pay for non-biotech food products. A third, small category includes several studies about the potential impacts of biofortified crops on public health in Asia.

Authors generally concur that Chinese consumers are more accepting of biotech food products than consumers in other countries. This finding is documented in both attitude surveys and estimated willingness to pay for non-biotech food, which is much lower in China than in other countries studied. Authors ascribe this finding to a combination of government policy and cultural and political history that is unique to China. A common thread throughout all of the consumer studies examined is the crucial role of information in changing the attitudes consumers express—and negative messages are especially potent.

Later econometric studies recognize the endogeneity of consumer awareness and willingness to pay, as well as purchase and payment decisions. Analysts identify and contrast the characteristics of consumers in market segments, pinpointing gender- and age-related differences. One problem for researchers is that the most widely adopted biotech crop is Bt cotton, which leads to an unfortunate disconnect between the farmer and consumer sides of the market. Since farmers will not have a market for their products if consumers do not accept them, in this next generation of studies it will be of fundamental importance to begin linking farm and consumer research in the same developing economy. The use of recommended approaches, such as those that combine revealed and stated preferences, will depend on whether or not biotech food products have been introduced.

Sector studies combine the two sides of the market and, in the case of biotech crops, also include the innovator: the fact that intellectual property rights confer a temporary monopoly on the supplier of the innovation is of particular interest. Monopolistic pricing behavior, expressed in high seed prices, can curtail benefits for both farmers and the innovator, as has been documented for the case of Bt cotton in Argentina. The purpose of sector studies is to provide investors and policymakers with estimates of the size of economic benefits and indicate how the "benefits pie" is shared among farmers, consumers, and innovators. With respect to biotech crops, most of the two dozen peer-reviewed articles that met the search criteria were ex ante. Thus authors aimed to provide preliminary information to guide strategic investment decisions, but, in a number of cases, the biotech crop they modeled has still not been released (such as biotech sweet potato and maize in Kenya and biotech potato in Mexico).

By far the most common approach in sector modeling is the economic surplus method, although the Global Trade Analysis Project (GTAP) model has also been applied to study the Chinese national economy, and one article deployed a linear programming model to estimate the potential impact of Bt cotton in West Africa. A small set of studies has begun to address the costs, but not the benefits, of regulation—and a stronger theoretical foundation is clearly needed. Aside from the adaptation of the economic surplus approach to incorporate monopoly pricing, a significant improvement in some studies has been the augmentation of the approach to include stochastic simulation. One recent study links the economic surplus model to a farm decisionmaking model in order to "endogenize" adoption.

Model parameters lead to wide variation in the size of potential benefits and their distribution among actors, although the magnitude of the transfer fees em-

bodied in the seed price and the time lag until release and adoption are generally shown to be critical determinants of these patterns. As with partial budgeting, the economic surplus approach is an essential tool for applied agricultural economists, and as such its shortcomings have been well documented. Economic surplus models portray partial equilibria, assume well-functioning markets, and have not yet incorporated positive and negative externalities, including environmental and health impacts. As in any model, the quality of the underlying data determines the validity of the findings. As noted in the discussion of farm impacts, data quality has in some cases been questionable. Time-series data in China and India are now emerging, and these data will enable much more reliable measurement of sector benefits.

Given how important supply channel performance and industrial organization are for biotech crops, the paucity of studies addressing these topics is surprising. There is much room for improvement during the next decade of research, because understanding these aspects of the industry will shed more light on whether the benefits projected will actually be earned.

Trade studies are conducted at a higher level of aggregation than sector studies: they assess the effects of the adoption of biotech crops in multiple sectors and multiple countries. The four major biotech crops on the market today (soybeans, cotton, maize, and canola) are all major internationally traded commodities. Trade studies of biotech crops differ from trade studies of other crops because they must account for specific trade-related regulations and the potential for market segregation. Varying degrees of consumer acceptance among trading nations raise some thorny policy issues.

The two dozen or so trade analyses that met search criteria fall into three groups. All are based on ex post trade data, but almost all are ex ante analyses. The first group of studies consists of descriptive accounting analyses of trade patterns. These analyses are not based on an explicit model of international markets. The other two represent two major categories of applied trade models: partial equilibrium models applied to one or more sectors in a few countries and multicountry, general equilibrium models. The strong structure of the partial equilibrium model, which also offers flexibility in investigating the effects of specific policies or institutional arrangements, is sacrificed for a high degree of aggregation in the general equilibrium model. The partial equilibrium models, like economic surplus models, do not take into account linkages with multiple sectors and markets.

The array of approaches nevertheless supports three main findings. First, they illustrate the economic advantage of being the first country to adopt: adopting countries both produce more and gain market share from non-adopters. Second, a number of studies demonstrate "immiserizing growth" when yield-enhancing technologies are adopted in markets with inelastic demand—benefiting consumers but not the adopting farmers. Third, studies suggest that the potential export losses from adopting biotech crops are outweighed by the potential gains from higher productivity. Fears that biotech-adopting and biotech-exporting countries will lose from market restrictions appear to have been overstated.

So far, these trade models remain unsophisticated, compared with sector models, with respect to treatment of risk and uncertainty. Assumptions needed to aggregate across nonhomogeneous countries, sectors, crops, and varieties can be distortive. One example is that the shift in the supply curve for biotech maize has been represented in a number of GTAP analyses by a shift in the entire cereal sector. Clearly, most models do not realistically represent developing-economy markets or the supply chain for biotech crops. Other than descriptive studies, no ex post study of the impact of transgenic crops on international markets was found.

A number of improvements are needed in the next decade of studies. Productivity changes induced by adoption of biotech crops should be more fully articulated in terms of regional differences, effects on labor markets, heterogeneous land types, and seed prices. Segmented markets, labeling effects in retail markets, and imperfections in input markets could be better addressed. Trade-related regulations and markets for unauthorized seed merit more attention.

Overall, the balance sheet remains promising for the few biotech crops that have been introduced in developing economies. This is especially true for emerging economies (such as China and India) with vibrant research institutions and strong markets and for commercially oriented farmers with good access to those markets and to knowledge about the technology. Still, the results from the few crops and traits studied, and the few in-depth country case studies, should by no means be generalized. The haste to generate information for an eager, awaiting public should be countered by more careful and more comprehensive research design.

Chapter 1

Introduction

A vast literature has accumulated since transgenic crop varieties were initially released to farmers in 1996. Several years after their introduction in the United States, crop varieties with transgenic resistance to insects or herbicide tolerance were supplied to farmers in countries with developing economies and nonindustrialized agriculture. Essays, editorials, newsletters, web conferences, articles, and books have argued the pros and cons of transgenic crops.[1] The global debate continues in this second decade of their use. A comparatively minor segment of this literature consists of studies conducted by agricultural economists to measure the impact of transgenic crop varieties on farmers, the size and distribution of the economic benefits from adopting them, consumer attitudes toward products made with transgenic ingredients, and implications of the use of transgenic crops for international trade. An even smaller subset treats the impacts of transgenic crops in developing economies.

Objectives of the Review

This food policy review summarizes that portion of the applied economics literature that deals with the impact of transgenic crops in nonindustrialized agriculture during the first decade of their adoption, with an emphasis on methods.

There are several reasons for our decision to focus on methods rather than findings. First, a number of studies have surveyed the findings thoroughly for both industrialized and nonindustrialized agriculture at several points in time. Surveys of methods are less common. So far each survey of methods has examined only one

[1] Reviews conducted by Frohlich (2005) and Fransen (2006) provide useful insights into the broader literature.

perspective or research question; this review assesses multiple research questions. A second reason is the recognition that the methods applied by researchers influence the nature and interpretation of their findings. Understanding the methods applied therefore enhances the comprehension of research findings. Third, our aim is to support researchers who seek to produce objective, relevant information about emerging crop biotechnologies for national policymakers in developing agricultural economies.

Scope of the Review

The number of references about biotechnology is immense even when limited by subject field. On July 31, 2007, we performed a search on CAB Direct using the query "economic* AND biotechnolog*". CAB Direct displayed a total of 9,823 records.[2] Google Scholar displayed 24,800 hits for a similar query, "biotechnology AND economics", on the same day.

We searched the English-, French-, and Spanish-language literature systematically, using a combination of web-based bibliographic databases, compilations of biotech-related literature that are available online, references cited in published articles, and email communications with economists who are currently implementing research. The four principal electronic bibliographic databases that have been consulted are CAB Direct, ISI Web of Science, EconLit, and Google Scholar.

For consistency in drawing comparisons among studies, the boundaries of the search were carefully delineated according to both general and specific criteria. General criteria include subject matter and review status. In terms of subject matter, only papers that examine the impacts of transgenic crops in nonindustrialized agriculture, including at least one developing economy, were considered. Purely theoretical studies, conceptual papers, and critical essays were omitted, as was literature about types of crop biotechnology other than transgenic crops, such as tissue culture.

Peer review was the second general criterion. Journal articles, book chapters, and published conference proceedings were included. Papers in progress and papers accepted for presentation at conferences have been excluded until notification is received of forthcoming publication. Technical reports, working papers, discussion papers, and reports published by private organizations, nongovernmental organizations, and local, state, or national governments have been included if there is some stated evidence that they have been reviewed. While acknowledging that there is both a range among peer-reviewed journals and a gradation of peer review among the publications we have included, establishing more restrictive criteria would sig-

[2]This query contains two wild cards. Wild cards identify all words that contain the truncated word as a subset.

nificantly reduce the amount of literature we can address. On the other hand, venturing into the non-peer-reviewed literature would inflate the number of publications and place us even more squarely in the realm of polemics.

Specific criteria were employed to distinguish review or survey papers from methods papers, and to differentiate among methods papers. Papers were classified as methods papers if authors stated an economic method or applied it to empirical data collected for the purpose of the study. These were organized according to four major research questions addressed by the literature. Each question corresponds to a component of the agricultural economy: (1) What are the (potential, actual) advantages of transgenic crops to *farmers?* (2) What are *consumers* willing to pay for products without transgenic ingredients? (3) What are the magnitude and distribution of the economic benefits from adoption of transgenic crops in an *industry (sector)?* (4) What is the international distribution of economic benefits from adoption and *trade* of transgenic crops?

Organization of the Review

In Chapter 2 we begin with tabular summaries of the search process. Reviews of findings and methods are listed by author, year, and topic area. Next we present numbers of empirically based articles by subject area, country, and crop-trait combination. Chapters 3–6 present a textual description of general findings with reference to methodology, in chronological order, by country case study. The concluding chapter summarizes the limitations of the studies published during the first 10 years of adoption and impact of transgenic crops in developing agriculture; it also recommends research directions for the next decade.

Chapter 2

Search Summary

Table 2.1 lists articles that review findings or methods. Of these 11 are global in coverage, only 4 focus on industrialized agriculture, and 12 address impacts in nonindustrialized agriculture. These numbers suggest a relatively high level of professional interest in the potential and actual impacts of transgenic crops in developing economies.

The count of articles in which authors have applied a stated economics method to an empirical dataset is shown in Table 2.2. A total of 137 articles published from 1996 through the end of 2007 met the criteria we established (see Chapter 1, "Scope of the Review"). Some of these 137 articles analyze more than one aspect. Overall, 67 examine impacts on farmers, 27 analyze consumer impacts, 27 treat sector (industry) impacts, and 27 assess impacts on international trade. Thus, as indicated by counts of peer-reviewed publications, evaluating the direct impacts of transgenic crops on farmers represents the foremost concern of applied economists during the first decade of adoption in nations with nonindustrialized agriculture.

Table 2.3 shows the count of articles by research question and crop-trait combination. By far the most researched crop-trait combination is insect resistant (IR) cotton (63 publications). Publications that analyze the impacts of transgenic maize, rice, and soybeans follow, with only 14–16 articles each. Twenty articles address transgenic crops in general. The same number of articles examines a range of other crops: bananas, cassava, coarse grains, eggplant, mustard, oilseeds, potatoes, sweet potatoes, and wheat. Categories total to more than the total number of articles (151 compared to 137) because some publications treat more than one crop-trait combination.

Table 2.4 reports the frequencies of publications by research question and country. As in Table 2.3, categories total to more than the total number of publications because some publications consider more than one country. Given that the

Table 2.1 Reviews of the impacts of transgenic crops in industrialized and nonindustrialized agriculture

First author	Year	Publication	Crops	Period	Focus	Review type
A. Global						
Brookes, G.	2006	*AgBioForum*	Multiple	1996–2005	Global	Findings
Babu, S. C.	2003	*Asian Biotechnology and Development Review*	Multiple	—	Global	Methods
Food and Agriculture Organization of the United Nations	2004	*State of Food and Agriculture 2003–2004*	Multiple	1996–2003	Global	Findings
Fernandez-Cornejo, J.	2006	Economic Research Service Economic Information Bulletin	Multiple	1996–2006	United States	Findings
Lusk, J. L.	2005	*Journal of Agricultural and Resource Economics*	Multiple	1992–2003	Global	Methods
Marra, M.	2002	*AgBioForum*	Multiple	1996–2002	United States	Findings
OECD	2000	Organisation for Economic Co-operation and Development Research Study	Multiple	—	Global	Methods
Purcell, J.	2004	*AgBioforum*	Cotton	1996–2003	Global	Findings
Scatasta, S.	2006	Mansholt Graduate School of Social Sciences Working Paper	Multiple	1997–2004	Global	Methods
Shoemaker, R.	2001	Economic Research Service Agriculture Information Bulletin	Multiple	1996–1999	United States	Findings
Wu, F	2004	RAND Corporation monograph	Multiple	1996–2003	Global	Findings

B. Industrialized countries						
Carpenter, J.	2001	National Center for Food and Agricultural Policy report	Cotton, maize, potato, soybeans	1996–2000	United States	Findings
Caswell, M. F.	1994	Economic Research Service Agricultural Economic Report		Up to 1994	United States	Findings
Demont, M.	1999	Department of Agricultural and Environmental Economics, Katholieke Universiteit Leuven, Working Paper	Multiple	—	European Union	Methods
Price, G.	2003	Economic Research Service Technical Bulletin	Cotton, soybean	1997	United States	Methods, findings
C. Nonindustrialized countries						
Falck-Zepeda, J.	2003	Organisation for Economic Co-operation and Development	Multiple	—	General	Findings
Fok, M.	2005	*Economie Rurale*	Cotton	1998–2002	China	Findings
Huesing, J.	2004	*AgBioforum*	Multiple	1996–2003	General	Findings
Qaim, M.	2005	*Quarterly Journal of International Agriculture*	Multiple	1996–2005	General	Findings
Raney, T.	2006	*Current Opinion in Biotech*	Cotton, maize, soybeans	1996–2005	General	Findings
Schaper, M.	2001	Fundación Futuro Latinoamericanopaper	Multiple	1996–2001	Latin America	Findings
Da Silveira, J.-M.	2005	Bellagio workshop paper	Soybean	1996–2004	Brazil	Findings
Toenniessen, G.	2003	*Current Opinion in Plant Biology*	Multiple	1996–2001	General	Findings
Nuffield Council on Bioethics	2003	Nuffield Council on Bioethics discussion paper	Banana, cotton, rice, soybeans, sweet potato	1998–2002	General	Findings
Trigo, E.	2002	Inter-American Development Bank technical paper	Multiple	—	LAC	Findings
	2006	ArgenBio study	Soybeans, cotton, maize	1996–2005	Argentina	Findings
Qaim, M.	1998	ZEF Discussion Paper on Development Policy	Multiple	—	General	Methods

Note: — indicates years not specified.

Table 2.2 Count of publications about the economic impact of transgenic crops in developing economies, by research question, 1996–2007

Research question	Number
Farmers	57
Farmers, industry	9
Farmers, trade	1
Consumers	26
Consumers, industry	1
Industry	17
Trade	26
Total	137

Notes: Selection criteria are described in Chapter 1. Number includes articles written in English, French, and Spanish.

Table 2.3 Count of publications about the economic impact of transgenic crops in developing economies, by research question and crop (trait)

Transgenic crops	Farm	Farmers/ industry	Farmers/ trade	Consumers	Consumers/ industry	Industry	Trade	Total
Cotton (IR)	46	3	—	—	—	8	6	63
Maize (IR)	5	1	1	—	—	4	3	14
Rice (HT, IR)	2	—	—	5	1	3	5	16
Soybeans (HT)	3	1	—	3	—	1	8	16
All other crops	1	4	—	6	—	5	6	22
General	—	—	—	16	—	1	3	20
Total	56	9	1	30	1	22	31	153

Notes: HT, herbicide-tolerant; IR, insect-resistant. — indicates no studies in this category. Other crops include bananas, cassava, coarse grains, eggplant, mustard, oilseeds, potatoes, sweet potatoes, and wheat.

subject of *Bacillus thuringiensis* (Bt) cotton dominates the total number of publications, the overall frequency distribution among categories and countries is heavily affected by the distribution among publications about Bt cotton. As a consequence, China, India, and South Africa are the most represented. Argentina is also well studied relative to other countries, and research on the situation in the Philippines is rapidly accumulating. The remaining countries are represented by five or fewer studies each.

Table 2.4 Count of publications about the economic impact of transgenic crops in developing economies, by research question and country

Country	Farmers	Farmers/ industry	Farmers/ trade	Consumers	Consumers/ industry	Industry	Trade	Total
China	11	1	—	13	—	1	7	33
India	19	2	—	4	—	3	1	29
South Africa	18	—	—	—	—	—	—	18
Argentina	5	1	—	1	—	1	2	10
Philippines	1	2	—	1	1	2	1	8
Mexico	1	2	—	—	—	1	1	5
West African countries[a]	—	—	—	—	—	3	1	4
Colombia	—	—	—	1	—	2	1	4
Indonesia	—	1	—	—	—	1	1	3
Kenya	—	1	—	—	—	1	—	2
Brazil	1	—	—	—	—	—	1	2
Vietnam	—	—	—	—	—	1	1	2
Taiwan	—	—	—	2	—	—	1	3
Bangladesh	—	—	—	1	—	—	1	2
Korea	—	—	—	1	—	—	1	2
Other countries[b]	1	—	1	3	—	2	2	9
Global	—	—	—	1	—	2	16	19
Total	57	10	1	28	1	20	38	154

Note: — indicates no studies in this category.
[a]Benin, Burkina Faso, Cameroon, Central African Republic, Chad, Côte d'Ivoire, Mali, Senegal, and Togo.
[b]Chile, Costa Rica, Egypt, Romania, South Korea, Thailand, Uganda, and Uruguay.

The preponderance of studies that document farm impacts, and especially of studies about Bt cotton, leads us to devote relatively more space to this research question and crop-trait combination in this report. Additional tables summarizing other details about publications—including authors, year of publication, and other descriptors—are provided in the chapters that follow.

Chapter 3

Impacts on Farmers

Approaches

By far the majority of studies about the adoption and impacts of transgenic crops on smallholder farmers examine the evidence (ex post) by analyzing primary data collected in personal interviews, farm records kept by companies, or trials supervised by researchers on farms. A few studies attempt to gauge the potential impact on farmers (ex ante), drawing on data collected from representative farmers whose characteristics indicate that they are likely future adopters of transgenic crops.

Authors who have documented impacts ex post use two main approaches, often applied to the same survey data. The first is farm accounting, or the use of partial budgets. Researchers compare the net profits per hectare for adopters and nonadopters, considering changes in costs and benefits that vary. They test a set of hypotheses that are considered to be fundamental to any economic evaluation of technologies: changes in yield, amounts and costs of inputs, and profits. For insect resistance conferred through Bt expression, the predominant trait examined so far, changes in amounts and costs of insecticide applied and changes in labor related to insecticide application are key parameters of interest. The second approach entails the specification of a statistical model that is grounded on a theoretical economics framework, such as a production function or a random utility model. These models test the same and related hypotheses more rigorously.

While rare, there are some ex ante predictions of farm impacts in the literature. In some cases authors apply similar econometric approaches to data from on-farm trials before the transgenic crop has been released to farmers (Qaim and Zilberman 2003; Huang et al. 2005). To assess the potential economic benefits to Chile of opening the domestic market to transgenic seed, Díaz Osorio et al. (2004) compare the production costs of transgenic (Bt and Roundup Ready [RR] maize) and nontransgenic maize by consulting survey data from 10 companies. Edmeades and Smale (2006) predict farmer demand for disease- and pest-resistant bananas in the East African highlands using a trait-based model and survey data that detail cultivar attributes and the characteristics of farmers, households, and markets. Their statistical model stems from the theory of the farm household, which considers the role of imperfect markets in production decisions. The authors chose this approach, rather than a model based on profit maximization, because banana is a semisubsistence crop in the region they studied.

Four recent articles (Kolady and Lesser 2006, 2007, 2008; Krishna and Qaim 2007) also explore ex ante the potential adoption and impact of IR eggplant in India. Considering the set of articles published from 1996 through 2007, these articles are remarkable in at least three respects. First, they are the first analyses of transgenic vegetable crops. Second, they address the implications of the decision by the Maharashtra Hybrid Seed Company (Mahyco) to donate the Bt gene royalty-free to public institutes in order to develop open-pollinated varieties (OPVs) of Bt. Related to this second feature is their explicit comparison of hybrids and OPVs as the host plants for the transgene. The authors deploy a range of techniques applied to farm and trial data—including contingent valuation, production functions, interval regressions, and sensitivity analysis—to contrast the characteristics of future OPV and hybrid adopters.

In a number of cases, including two of the eggplant studies, authors use partial budgets as inputs for the assessment of the potential impact of transgenic crops on the sector (including farmers, consumers, and innovators). Sector analyses, most frequently derived from an economic surplus model, are reviewed in Chapter 5.

A unique approach by Birol, Villalba, and Smale (2007) examines farmers' preferences as consumers of seed. Data were collected from 420 farmers in the states of Jalisco, Michoacán, and Oaxaca with a choice experiment survey. The authors use a latent class model to characterize Mexican farmers in terms of their propensity to continue *milpa* production and their need for compensation if transgenic maize were introduced.[1]

We now work through the literature that examines impacts ex post in chronological order, clustered by crop-trait combination and country. Reviewing the literature by chronology provides a sense of where a new method or dataset sheds new light on a previous finding. The most extensive body of ex post evidence has been compiled for IR cotton, which leads us to devote more of the chapter to this crop-

[1] *Milpa* is an ancient, complex intercropping system based on maize, beans, and squash that now assumes multiple forms.

trait combination. Major country case studies for IR cotton are represented by sets of studies conducted in China (14), South Africa (13), and India (11), with a few articles each on Argentina (3) and Mexico (2). Several articles trace the impacts of herbicide-tolerant (HT) soybeans on farmers in Argentina and Brazil, and IR maize in Chile, Mexico, the Philippines, and South Africa. These are grouped together in a section that precedes the chapter conclusions.

IR Cotton

South Africa

Of the 15 articles that meet our criteria and address the impacts of Bt cotton on smallholder farmers in South Africa (Table 3.1), 7 are based solely on analysis of data collected from the same sample of 100 farmers in Makhathini Flats, a development scheme in KwaZulu, Natal. As can be expected when adoption of a new technology is in its initial phases, the initial research in Makhathini Flats was purposely placed. Thus there may be statistical bias associated with study placement, and findings can be generalized to other locations only to the extent that these locations share comparable characteristics.

Authors recognized the potential for placement bias from the outset. Thirtle et al. (2003, 731) describe Makhathini Flats as "a low potential area for cotton production" and "atypical in that the biotech companies are locally present and support services are unusually good, which affects the wider applicability of this study." Over 31,500 hectares were planted to cotton in South Africa in 2001/02, with 22,000 hectares in the drylands, of which Makhathini Flats represented only 31 percent (6,800 hectares) (Gouse, Kirsten, and Jenkins 2003). In 2002–04, insecticide prices were substantially higher in Makhathini Flats than in other parts of the province (Hofs, Fok, and Vaissayre 2006).

Authors also mention their concerns about sample selection bias. For example, "there was some potential for bias in the selection process, as Vunisa [Cotton] agents purposely targeted farmers with larger areas of cotton during the first year of Bt cotton release, and to a lesser extent also in the second year" (Ismael, Bennett, and Morse 2002a, 3).

In general, wide swings in climatic conditions also mean that findings from a farmer survey in any one year cannot be construed as representative across seasons. Though the original survey spanned two seasons (1998/99 and 1999/2000), neither year was normal; there was drought in the first season and late, heavy rains in the second (Kirsten and Gouse 2003). Reflecting this situation, Ismael, Bennett, and Morse (2002a, 2002b) estimated that gross margins for Bt cotton were 11 percent higher than for non-Bt cotton in 1998/99 and 77 percent higher in 1999/2000.

Table 3.1 Study descriptors: Impacts of Bt cotton on farms in South Africa

Authors	Year	Data type	Sample size	Methods
1. Bennett, R., S. Morse, and Y. Ismael	2006	Farm records	1,283; 441; 499; 32; 100	Farm record analysis, farm survey analysis, Gini coefficient
2. Bennett, R., T. Buthelezi, Y. Ismael, and S. Morse	2003	Unclear	32	Case study interview
3. Bennett, R., Y. Ismael, S. Morse, and B. Shankar	2004b	Company data	1,283; 441; 499	Farm record analysis, production function, Gini coefficient, biocide index
4. Gouse, M., C. Pray, and D. Schimmelpfennig	2004	Statistical survey, key informant	143 (100 smallholders; 43 large-scale farmers)	Farm survey analysis
5. Gouse, M., J. Kirsten, and L. Jenkins	2003	Statistical survey	Unclear	Farm survey analysis, data envelope analysis, value of marginal product curve
6. Gouse, M., J. Kirsten, B. Shankar, and C. Thirtle	2005a	Statistical survey, key informant	100	Farm survey analysis, stochastic production frontier, damage control production function, value of marginal product analysis, institutional analysis
7. Hofs, J.-L., M. Fok, and M. Vaissayre	2006	Statistical survey	20	Farm survey analysis, plot monitoring
8. Ismael, Y., L. Beyers, C. Thirtle, and J. Piesse	2002	Statistical survey, key informant	100	Farm survey analysis, adoption model, stochastic production frontier, deterministic frontier programming model, Gini coefficient
9. Ismael, Y., R. Bennett, and S. Morse	2002a	Statistical survey	100	Farm survey analysis
10. Ismael, Y., R. Bennett, and S. Morse	2002b	Statistical survey	100	Farm survey analysis
11. Ismael, Y., R. Bennett, and S. Morse	2001	Statistical survey	100	Farmer survey analysis, partial budget
12. Morse, S., R. Bennett, and Y. Ismael	2006	Company data, sample survey data	2,200	Farm survey analysis, biocide index, environmental impact quotient
13. Morse, S., R. Bennett, and Y. Ismael	2005a	Company data	1,283; 441; 499	Farm record analysis
14. Shankar, B., and C. Thirtle	2005	Statistical survey	100	Farm survey analysis, damage control production function, tests for endogeneity of pesticide use and Bt choice, model tests, value of marginal product analysis
15. Thirtle, C., L. Beyers, Y. Ismael, and J. Piesse	2003	Statistical survey, key informant	100	Farm survey analysis, adoption model, stochastic efficiency frontier

Authors also use larger samples of farm records assembled by the Vunisa Cotton company. Critics often discount the validity of company records by arguing that they are likely to be biased in favor of the technology, despite the larger sample size. On the basis of farm records drawn over three growing seasons (1998/99 to 2000/1), Morse, Bennett, and Ismael (2005a) concluded that farmers in Makhathini Flats benefited on average from higher yields, lower costs for spraying insecticides, and higher gross margins.

Establishing positive effects of adoption on average farm profits is fundamental. A second major hypothesis concerns differential impacts according to farm size, which are an indicator of unequal capital endowments and access to farming resources. Analyses that compare economic returns per hectare between smallholder and larger-scale producers in South Africa conclude that smallholders are major beneficiaries of Bt cotton (Ismael, Bennett, and Morse 2002b; Gouse, Kirsten, and Jenkins 2003). Gouse, Pray, and Schimmelpfennig (2004) surveyed large-scale farmers in irrigated areas of the North Cape and Mpumalanga, as well as dryland farmers in the Springbok flats, Limpopo Province, combining these data with those of the 1999/2000 sample from Makhathini Flats. They find that, on average, large-scale farmers in irrigated areas earn the greatest yield benefits per hectare, the greatest reduction in pesticides applied, and the largest income advantage in terms of amounts. Small-scale farmers gain the highest yield increase of the three groups in terms of percentages. Small-scale farmers also earn a larger income advantage than larger-scale farmers in dryland areas, but they benefit least from reduction in pesticide use. Gouse, Pray, and Schimmelpfennig (2003, 2004) report that larger-scale farmers save in terms of lower diesel and tractor costs and in terms of "managerial freedom." While there is some evidence that Bt cotton reduced inequality in Makhathini Flats, Ismael et al. (2002, 346) concluded that "the per capita distribution of income from cotton in this area is about as unequal as the distribution of per capita incomes in the Western European countries."

Researchers in South Africa also hypothesized that a major attraction of Bt cotton would be the possibility of saving labor. Makhathini Flats is an area that has been burdened with HIV/AIDS. The duress involved in back-spraying and collecting water for spraying (often accomplished by women and children) cannot be overstated. Often, however, labor costs were not recorded in farm budgets. Kirsten and Gouse (2003) note that the labor saved by fewer applications of pesticides could have been canceled out by the increased labor needed to harvest more output. In their most recent published work, Shankar and Thirtle (2005) conclude that Bt was not labor-saving in their sample of smallholder farmers.

In the South Africa case, researchers have tested more subtle hypotheses over time with increasingly sophisticated econometric approaches. The initial approaches included deterministic frontier models (Ismael et al. 2002), stochastic frontier models (Ismael, Bennett, and Morse 2002b; Thirtle et al. 2003), and data envelope analysis (Gouse, Kirsten, and Jenkins 2003). Gouse, Kirsten, and Jenkins (2003) and Thirtle et al. (2003) found that Bt cotton growers, whether smallholders or large-scale farmers, were more technically efficient than growers of non-Bt cotton. Gouse et al. (2005a) subsequently estimated a production function with damage abatement. This approach considers pesticides, unlike inputs such as fertilizers, to be damage-abating as well as output-enhancing.

In the most thorough analysis based on the survey of 100 farmers in Makhathini Flats, Shankar and Thirtle (2005) also estimated this type of production function. They explored the efficiency of pesticide use by comparison with the estimated value of the marginal product. In addition the authors explicitly tested for sample selection bias and for the endogeneity of pesticide use. If pesticide use were endogenous, the same factors that influence yield would also influence whether or not the farmer chooses to apply pesticides, leading to biased regression coefficients. Shankar and Thirtle (2005) conclude that farmers do not apply pesticides in response to pests but in a predetermined, prophylactic way.

Pesticides are not heavily used by smallholders in Makhathini Flats, while they are overused in Argentina and China. Yields in the sample are 600 kilograms per hectare, as compared to 3,000 kilograms per hectare in China. Unlike the models they used in earlier studies (data envelope analysis, stochastic frontier), the production model with damage abatement shows that pesticide use has a positive effect on cotton productivity and that smallholders underuse pesticide relative to the economic optimum. This holds true whether they grow Bt or non-Bt cotton. Nonetheless their analysis confirms that in this group of smallholders, during the 1999/2000 season, the effect of Bt cotton on yield was more important than its effect on abating damage.

A key conclusion reached by Shankar and Thirtle (2005) is that adoption in Makhathini Flats was driven by supply rather than by demand. When adoption is driven by supply, whether a new variety fails or succeeds is particularly sensitive to the organization of the marketing channel, a point underscored by Gouse et al. (2005a). Over 90 percent of cotton farmers in Makhathini Flats grew Bt cotton in 2001/02. The Vunisa Cotton company supplied growers with inputs and credit and bought the cotton they produced, also providing some extension advice. After a few seasons, farmers defaulted on loans from Vunisa by selling to a new gin, and in the following year no seed or credit was supplied. Production declined in subsequent seasons. Gouse, Kirsten, and Jenkins (2003) proposed that, contrary to expectations, it may have been the vertical integration in the cotton industry, with the monopsony of the local ginnery that also supplied seed and credit, which laid the foundation for success in Makhathini Flats. Vulnerability to external market arrangements compounds the challenges of a harsh production environment. For this reason,

Hofs, Fok, and Vaissayre (2006) caution that the income generated from growing Bt cotton may not be sufficient to generate tangible and sustainable improvement in farmer well-being. Morse, Bennett, and Ismael (2006) also note the continued reliance of growers in Makhathini Flats on a single company for inputs and sales.

Whether or not growing Bt cotton is associated with reduced pesticide use constitutes a major research hypothesis with broad-ranging implications for development policy. Reduced pesticide use can lead not only to lower production costs and labor savings, but also to lower exposure of farmers and the environment to hazardous chemicals. Bennett, Ismael, and Morse (2005) initially found that overall biocide indexes rose in Makhathini Flats with the introduction of Bt cotton. Bt growers applied lower amounts of pesticides and had lower biocide indexes than growers of non-Bt cotton. The authors note that Bt growers reduced both non-bollworm and bollworm insecticides, perhaps due to a misunderstanding about the technology (Bennett, Ismael, and Morse 2005; Morse, Bennett, and Ismael 2005a).

Hofs, Fok, and Vaissayre (2006) compared near-isogenic lines and monitored the practices of a different sample of 20 farmers in Makhathini Flats. They observed a decrease in pyrethroid use during the 2002/03 and 2003/04 seasons, though farmers did not abandon the pesticide. At the same time, farmers applied substantial amounts of organophosphates to control pests not affected by the Bt toxin. The extent of the labor savings was not as great as expected. Surprisingly, "more money was invested in insect management for Bt cotton than for crops other than Bt cotton, probably because farmers . . . upgraded their seed-cotton yield objectives and adjusted their investment" (Hofs, Fok, and Vaissayre 2006, 5). This study illustrates the trade-offs that occur when rigor is pursued in a multidisciplinary analysis. While concern with identifying the correct counterfactual led the authors to use isogenic lines, and daily records improved the quality of data concerning farmers' practices, only 20 farmers were studied and these were located in close proximity to one another.

Two 2006 publications review impacts of Bt cotton on profitability and the environment from 1998/99 by drawing on extensive farm record and survey data. Bennett, Morse, and Ismael (2006) conclude that while adoption is associated with slightly larger farm sizes in years 1 and 3, adopters are more likely to have smaller farm sizes in year 2. In all three seasons, adopters had advantages over nonadopters in terms of average gross margins, but this was particularly the case in the wetter year. They report data suggesting that the number of accidental pesticide poisoning cases has declined.

After quantities of insecticide applied by farmers were converted into a biocide index and an environmental impact quotient (EIQ), to allow for differences in toxicity and persistence in the environment, Morse, Bennett, and Ismael (2006) found evidence of environmental benefits associated with growing Bt cotton, for all three seasons of study and for both bollworm and non-bollworm categories of insecticide. They note that toxin expression varies significantly across seasons, by cultivar, and by plant part.

Morse, Bennett, and Ismael (2006) discuss the pros and cons of measuring the environmental impact of pesticides with either the biocide index or the EIQ. The biocide index is defined as the sum of the "toxic load" of pesticides used over the duration of the land use system. Toxic load is a function of the amount of a pesticide applied each year, the concentration and toxicity of the active ingredient, and how long it remains active in the environment. One of the problems with the biocide index is its sole emphasis on mammalian toxicity. The EIQ was developed to score insecticides for chronic and dermal toxicity to mammals, persistence on plant material, and toxicity to fish, birds, and beneficial arthropods. Insecticides differ in terms of their persistence in the environment and their toxicity to mammals, fish, and beneficial insects (including natural enemies of cotton pests). All methods for computing these indexes are founded on assumptions about what to include in the equations and how to calculate the formulas.

China

Huang and colleagues have implemented continuous, in-depth survey research since 1999 (Pray et al. 2001; Huang et al. 2002a, 2002b, 2002c, 2003, 2004) (Table 3.2). Representing a coherent whole, these studies suggest that China may be the most successful case for Bt cotton in terms of sustained, widespread, and positive effects on farm profits; reduced pesticide use; health; and the environment. Nevertheless other authors have raised questions regarding the effectiveness of Bt and regional variation in the benefits earned by farmers (Xu et al. 2004; Fok et al. 2005; Yang et al. 2005; Pemsl, Waibel, and Gutierrez 2005).

Sample sizes for farmer surveys in China are large relative to those in South Africa. The first year of survey data in China (1999) included 282 farmers in Hebei and Shandong provinces. Yields did not differ significantly between Bt and non-Bt growers, but non-Bt growers used five times as much pesticide and paid seven times as much for it. Net income was positive for Bt growers and negative for non-Bt growers. Returns to labor were over twice as high for Bt growers.

Multivariate analysis of the first-year survey data, published in 2003, confirmed that Bt use reduced the utilization of pesticides, particularly organophosphates. Farmers benefited most from savings in pesticide expenditures and labor, since at that time the yields of major Bt and non-Bt varieties were statistically "indistinguishable" (Huang et al. 2003, 61). Authors reported that all Bt cotton varieties—including those introduced by foreign life science companies and those bred by China's research system—were equally effective.

Table 3.2 Study descriptors: Impacts of Bt cotton on farms in China

Authors	Year	Data type	Sample size	Methods
1. Huang, J., R. Hu, C. E. Pray, F. Qiao, and S. Rozelle	2003	Statistical survey	282	Farm survey analysis, multivariate pesticide use model (ordinary least squares)
2. Fan, C., J. Li, R. Hu, and C. Zhang	2002	Statistical survey	1055	Farm survey analysis
3. Huang, J., R. Hu, C. E. Fan, C. Pray, and S. Rozelle	2002a	Statistical survey, key informant	282; 407; 366	Farm survey analysis, yield model, pesticide use model, instrumental variable estimation, two-stage least squares, Cobb-Douglas function, damage control function
4. Huang, J., R. Hu, C. E. Pray, and S. Rozelle	2005b	Statistical survey, field trials (rice)	337/45; 494/122; 542/179; 123/224	Farm survey analysis, yield pesticide use model, instrumental variable estimation, two-stage least squares, Cobb-Douglas function, damage control function
5. Huang, J., R. Hu, Q. Wang, J. Keeley, and J. Falck-Zepeda	2002c	Statistical survey, key informant, trial data	282	Laboratory survey, farm survey analysis
6. Huang, J., R. Hu, S. Rozelle, F. Qiao, and C. Pray	2002b	Statistical survey, key informant, government data	282	Farm survey analysis, instrumental variable estimation, damage control production function, pesticide use model
7. Kuosmanen, T., D. Pemsl, and J. Wesseler	2006	Statistical survey, leaf tissue	150	Damage control production function (two-stage, semiparametric, interaction between pest exposure and damage control, plot monitoring, leaf tissue analysis)
8. Pemsl, D., H. Waibel, and A. P. Gutierrez	2005	Statistical survey, leaf tissue	150	Damage control production function, plot monitoring, leaf tissue analysis
9. Pray, C. E., D. Ma, J. Huang, and F. Qiao	2001	Statistical survey, key informant	283	Farm survey analysis, economic surplus model
10. Xu, J., Z. You, W. Wang, and Y. Yang	2004	Survey (possibly statistical)	—	Farm survey analysis, partial budget
11. Yang, P. Y., M. Iles, S. Yan, and F. Jolliffe	2005	Statistical survey	92	Farm survey analysis

Note: — indicates sample size not specified.

Huang et al. (2002b) then estimated a production function with damage abatement. They implemented an instrumental variables regression and specified interactions between use of Bt and use of pesticides. Findings regarding the effects of Bt cotton on efficiency and reduced use of pesticides were substantiated in this article. Econometric models demonstrated that Chinese farmers tend to overuse pesticides relative to the economic optimum, and field observation confirmed that they do not protect themselves. Thus the health benefits and reduced costs of Bt cotton are experienced immediately by farmers.

In one of their most thorough analyses, relying on three years of survey data and expanded sample coverage, Huang et al. (2002a) estimated a production function with damage control, but they also attempted to correct for the potential bias from endogeneity related to both pesticide use and variety choice. They concluded that (1) growing Bt cotton varieties *does* have a positive effect on crop yield; (2) Bt cotton reduces yield losses through abated damage; (3) pesticide use on non-Bt cotton varieties only abates damage; (4) benefits from Bt cotton vary across provinces, and are lowest in Henan and Jiangsu; and (5) farmers overuse pesticides, even when they grow Bt cotton.

The first conclusion reflects the fact that when comparisons are made without the use of isogenic lines, observed yield advantages are the outcome of the effectiveness of the trait, the genotype, management, environment, and interactions among all these factors. Trade-offs in yield potential and resistance levels among non-Bt cotton varieties, combined with the variety choices farmers make and their management practices, provide possible explanations for their results. The authors note that farmers generally grow non-Bt varieties that are resistant but lower in yield. Higher-yielding, more susceptible non-Bt varieties are grown on minor areas. On the other hand, once Bt substitutes for other mechanisms of genetic resistance, it is likely that farmers choose to grow the highest-yielding Bt varieties. Breeders are also likely to have inserted the gene into higher-yielding, susceptible varieties. Farmers who choose to grow Bt varieties may also be those who attain higher average yields.

As would be expected due to geographical variations in pest pressures, variety performance, and farmers' practices, the literature suggests some regional disparities in farm impacts. Fok et al. (2005) affirm the success of Bt cotton in the Yellow River region of China. According to the authors, resistance to insecticides had evolved in this region, and farmers applied 10–12 treatments, compared to a norm (in most countries) of 2–4 treatments. They also cite evidence that Bt cotton is not as successful in the Yangtze River Valley (Jiangsu) and other provinces, where pest pressures are lower and the germplasm is less well adapted.

Also in Jiangsu, Xu et al. (2004) found that although planting Bt cotton decreased the cost of insecticide and labor, it also decreased net benefits because yields were lower than for conventional cotton while the seed price was higher. Yang et al.

(2005) concluded that in Liqing County, Shandong Province, farmers grew more than six varieties of Bt cotton but were still overusing pesticides.

Pemsl, Waibel, and Gutierrez (2005) applied an exemplary protocol to measure and test the effect of adoption of Bt cotton on insecticide use and damage abatement in Shandong Province during the 2002 cropping season. In addition to frequent monitoring of insecticide use practices, they sampled leaf tissue. Instead of a zero-one variable to represent use of a Bt cotton variety (as in most other studies), the authors employed a continuous variable that measured the concentration of Bt toxin. To construct the variable, terminal leaves were collected from five different points for three plants in a row. The authors estimated a production function with damage abatement, modeled simultaneously with an insecticide use function. Their results confirmed that both Bt and non-Bt growers overused pesticides. Surprisingly, and contrary to other results, the econometric results showed that neither insecticide use nor Bt use reduced damage from bollworm. Pemsl and colleagues then tested the relationship between seed price and the effectiveness of Bt cotton seed. They concluded that almost 60 percent of the plots had cotton with levels of toxins that were lower than the standard, and while the proportion was higher for lower-priced seed or seed that had been multiplied on the farm, some higher-priced seed also had substandard concentrations of Bt toxin. In general, the variation in Bt expression was large.

India

Studies conducted in India (Table 3.3) illustrate several points of major importance for measuring the impact of Bt cotton on farmers. The first is that the more heterogeneous the growing environment, pest pressures, farmer practices, and social context, the more variable the benefits are likely to be. This truism holds for any new crop variety, no matter how widely adopted.

The physical, social, and economic environment for Bt cotton in India is indeed heterogeneous. Cotton is grown in most of India's agroecological zones on approximately 9 million hectares distributed over nine states. Sixty percent of this area is rainfed. While the most damaging pests are bollworms, hundreds of other pests are widespread. Soil and climatic conditions are difficult.

Several studies illustrate this point, with different approaches. Narayanamoorthy and Kalamkar (2006) collected data from 150 farmers during the 2003 rainy season in two districts in the Vidarbha region of Maharashtra. The authors targeted their analysis to pairwise yield comparisons among Bt and non-Bt hybrids (Mech 162 and Mech 184 for Bt; Bunny 145 and Ankur 651 for non-Bt). Overall, Mech 162 performed much better than Mech 184. Relative advantages of Bt over non-Bt differed between districts for the same hybrid and within districts by hybrid.

Bennett et al. (2004a) and Morse, Bennett, and Ismael (2005b) analyzed farm survey data for over 9,000 cotton plots. Gross margins per hectare were

Table 3.3 Study descriptors: Impacts of Bt cotton on farms in India

Authors	Year	Data type	Sample size	Methods
1. Bennett, R., U. Kambhampati, S. Morse, and Y. Ismael	2006	Statistical survey	7,751; 1,580	Farm survey analysis, production function
2. Bennett, R., Y. Ismael, and S. Morse	2005	Statistical survey	622	Farm survey analysis, multiple regression analysis
3. Bennett, R., Y. Ismael, U. Kambhampati, and S. Morse	2004a	Statistical survey	7,751; 1,580	Farm survey analysis
4. Crost, B., B. Shankar, R. Bennett, and S. Morse	2007	Statistical survey	338 farmers; 718 plots	Farm survey analysis, fixed effects, panel data, selectivity bias, Cobb-Douglas production function
5. Morse, S., R. M. Bennett, and Y. Ismael	2007	Survey (possibly statistical)	450	Farm survey analysis, Gini coefficient
6. Morse, S., R. M. Bennett, and Y. Ismael	2005c	Statistical survey	7,751; 1,580	Farm survey analysis
7. Morse, S., R. M. Bennett, and Y. Ismael	2005b	Statistical survey	622	Farm survey analysis
8. Naik, G., M. Qaim, A. Subramanian, and D. Zilberman	2005	Statistical survey	341	Farm survey analysis, production function
9. Narayanamoorthy, A., and S. S. Kalamkar	2006	Statistical survey	150	Farm survey analysis, production function
10. Pemsl, D., H. Waibel, and J. Orphal	2004	Statistical survey, literature	100	Stochastic partial-budget
11. Qaim, M.	2003	Sample from on-farm trials	157	Trial data analysis
12. Qaim, M., A. Subramanian, G. Naik, and D. Zilberman	2006	Statistical survey	341	Farm survey analysis, production function
13. Qaim, M., and D. Zilberman	2003	On-farm trials	157	Trial data analysis, yield-density function, logistic damage control function
14. Qayum, A., and K. Sakkhari	2006	Statistical survey, key informant, video, focus group, monitoring	440	Farm survey analysis
15. Sahai, S., and S. Rehman	2003	Survey (possibly statistical), key informant (traders)	100	Farm survey analysis, partial budget
16. Sahai, S., and S. Rehman	2004	Survey (possibly statistical)	136	Farm survey analysis, key informant

higher on Bt plots, but the difference was much greater in 2003 than in 2002, varying spatially among subregions. Employing a large sample of pooled cross-sectional and time-series data recorded at the plot level (collected by company extension agents), Bennett et al. (2006) estimated a production function that introduced use of Bt hybrids as both an intercept shift and an interaction variable. Their findings confirmed the spatial and temporal variation in partial productivity of Bt cotton. In some areas of Maharashtra State, they found that farmers did not benefit at all.

The economics of Bt cotton are determined by the severity of pressure from lepidopteran insects. To capture this element in the temporal and spatial variability of economic returns, Pemsl, Waibel, and Orphal (2004) performed a stochastic budget analysis. When risk and uncertainty were considered, they concluded that a prophylactic chemical control strategy would be superior to the use of Bt hybrids in both irrigated and nonirrigated cotton in Karnataka.

Naik et al. (2005) and Qaim et al. (2006) estimated a cotton production function for 341 farmers surveyed during the 2002 season in Andhra Pradesh, Karnataka, Maharashtra, and Tamil Nadu. Their analysis confirms the heterogeneity among farmers in terms of agroecological, social, and economic conditions: "Positive mean values for the technology yield or gross margin effects do not imply that every single farmer benefits. . . . a uniform experience with a new technology can hardly be expected" (Naik et al. 2005, 1515).

A second theme is unique to the India country case relative to others. An active civil society that is vocal for and against genetically engineered seed appears to have polarized the perspectives of both researchers and policymakers. Fervent discussion of methods and findings has been articulated in both the popular and the peer-reviewed literature. A broader range of opinion has in turn led to an improved understanding.

For example, Qaim and Zilberman based their optimistic estimate of the yield advantages of Bt cotton hybrids (80–87 percent) on data from on-farm trials of the first three approved hybrids in Madhya Pradesh, Maharashtra, and Tamil Nadu (Qaim 2003; Qaim and Zilberman 2003). Generally trial data are not considered representative of farmers' conditions, a fact that was acknowledged by Qaim. Arunachalam and Ravi (2003) and Sahai and Rehman (2003) were among the first critics of Qaim's results. Arunachalam and Ravi (2003) questioned the data, claiming that more reliable data from trials conducted by Punjab Agricultural University in 2002 showed that yields were higher for non-Bt germplasm than for the three hybrids released by Mahyco-Monsanto Biotech Ltd. (MMB).

Sahai and Rehman (2003) conducted a random sample survey of 100 farmers in Maharashtra and Andhra Pradesh during the first growing season after the commercial release of the Bt cotton hybrids. They compared the Bt hybrids Mech

184 and Mech 162 to local hybrids, Brahma and Banny.[2] Net profit from Bt cotton was lower per acre in all types of fields. The Bt hybrids had shorter duration but less vigorous growth, and they protected against green bollworm but not pink bollworm. Seed of Bt hybrids was four times more expensive than that of non-Bt hybrids. In 2004 the same authors implemented another survey in four districts of Andhra Pradesh, reporting economic losses for 60 percent of farmers growing Bt cotton hybrids. They argued that farmers sought unapproved Bt variants and good local hybrids because these outperformed the approved hybrids.

Mech 162, Mech 184, and Mech 12 were approved for release in March 2002 and sold in the states of Andhra Pradesh, Gujarat, Karnataka, Madhya Pradesh, Maharashtra, and Tamil Nadu. Despite poor rainfall and low pressure from the bollworm complex in that season, Barwale et al. (2004) reported an average 30 percent yield advantage for Bt hybrids compared to non-Bt hybrids, higher net profits, lower application rates for pesticides, and better cotton quality. Although the authors do not detail the methods they used to select their sample, the survey of 1,069 farmers was implemented by Mahyco extension workers in the six states. The confidence intervals reported for yields are wide, illustrating the variability in farmers' conditions (Barwale et al. 2004, 25, Table 3). Profits were calculated with imputed prices rather than survey data.

Seeking resolution in this debate led to recognition of the importance of not only the local environment but also host germplasm for the yield advantages attained by Bt growers. Considering the host germplasm is crucial for establishing the correct counterfactual. Authors suggested that the host germplasm of the first MMB hybrids was not broadly adapted to Indian growing conditions (e.g., Arunachalam and Ravi 2003; Sahai and Rehman 2004). Naik et al. (2005) and Qaim et al. (2006) emphasize that the better adaptation of local non-Bt hybrids compared to Bt hybrids (the germplasm effect) influences relative profitability.

Thus local adaptation of the germplasm into which the gene construct is backcrossed is critical to the success of the new transgenic seed. Concurring with this point, Bennett, Ismael, and Morse (2005) showed that official Bt varieties significantly outperform the unofficial varieties, but that unofficial, locally produced Bt hybrids can also perform better than non-Bt hybrids. They reported that second-generation Bt seed appears to have no yield advantage over non-Bt hybrids but can save on insecticide use. The Bt gene conferred some advantages even in the second generation of use, and farmers regarded it as transgenic seed.

Morse, Bennett, and Ismael (2007) appear to be the first to focus on the issue of inequality and adoption, using the example of Bt cotton in India. Data were

[2]Presumably this is the same variety as Bunny, spelled differently.

collected by recall for the 2002 and 2003 seasons from a sample of 450 farmers in two taluks (administrative divisions) of Jalgaon District in Maharashtra. The authors interpreted the gross margins per acre with Gini coefficients. They found that adoption of Bt hybrids enhances equality among the adopting farmers but reinforces the gap between adopters and nonadopters (including both Bt and non-Bt). This is explained by the fact that adopters of Bt hybrids also produced higher yields and revenues per acre in their non-Bt cotton, preferentially planting Bunny, which performs well in the particular growing environment.

This finding has relevance for the second analysis published by these authors in 2007. Crost et al. (2007) are perhaps the first to examine in detail the problem of selection bias when analyzing the impacts of transgenic crops on smallholder farmers. Their procedure corrects for the bias that is generated by the decision of more efficient farmers (who are also Bunny users) to adopt Bt cotton. After estimating a yield response function with panel data and a fixed-effects model, they still find a significant, positive yield effect from adopting Bt cotton.[3] Nonetheless, estimates of the mean yield advantage are more than twice as high when selection bias is ignored.

Argentina and Mexico

The country cases of Argentina and Mexico are less studied by far, but they offer insights into the effects of intellectual property rights (IPRs) on the economic benefits earned by farmers (Table 3.4). In Mexico IPRs were strictly enforced, as they are in the United States. To protect their revenue, Monsanto established contracts with farmers and gin owners. Farmers who desired access to the Bt cotton technology were obligated to forfeit the right to save seed and were required to have cotton ginned only where authorized. In their contracts farmers specified the total area to be planted, and Monsanto spot-checked cotton fields for compliance. Gins could become authorized (hence become monopsonists) by agreeing to refrain from selling Bt seed obtained during the ginning process. Contracts with Monsanto/Deltapine were drawn up to protect IPRs, and other agreements, with private-sector credit agencies, banks, and large cooperatives, allowed farmers access to credit. These contracts delineated the terms for technical assistance, the granting of credit, production, and product marketing. Given this institutional context, Traxler et al. (2003) and Traxler and Godoy-Avila (2004) concluded that Bt cotton

[3]Huang et al. (2005) assessed the potential impact of IR rice in China, using an experimental design to randomize the sample of on-farm trials. However, the potential for bias due to placement is not addressed through use of this approach, since trial locations are purposely selected. Since inputs are controlled and trials are managed by researchers, systematic bias in measuring the advantages of the new technology is also likely.

Table 3.4 Study descriptors: Impacts of Bt cotton on farms in Argentina and Mexico

Authors	Year	Data type	Sample size	Methods
A. Argentina				
1. Qaim, M., and A. de Janvry	2003	Statistical survey	299	Farm survey analysis, contingent valuation and contingent behavior, data enrichment, combined stated and preferences, adoption model and willingness to pay model
2. Qaim, M., and A. de Janvry	2005	Statistical survey, entomological and agroecological data for physiological model	299	Farm survey analysis, other (insecticide use and insecticide reduction functions), damage control production function (instrumental variable insecticide use model), simulation of physiological model of resistance, benefits by farm size
3. Qaim, M., E. J. Cap, and A. de Janvry	2003	Statistical survey, entomological and agroecological data for physiological model	299	Farm survey analysis, other (insecticide use and insecticide reduction functions), damage control production function (instrumental variable insecticide use model), simulation of physiological model of resistance, brief
B. Mexico				
1. Traxler, G., and S. Godoy-Avila	2004	Statistical survey, key informant	152; 242	Farm survey analysis, economic surplus, small open economy, brief
2. Traxler, G., S. Godoy-Ávila, J. Falck-Zepeda, and J. J. Espinoza-Arellano	2003	Statistical survey, key informant	152; 242	Farm survey analysis, economic surplus, brief

solved a major production problem for farmers in the Comarca Lagunera region of Coahuila and Durango states, where Bt is effective against the major pests. From a methodological perspective, this situation also meant that a moderate-size sample of farmers was judged to be representative.

In Argentina Monsanto strictly enforced IPRs on Bt cotton, contributing to low net returns and low rates of adoption (Trigo et al. 2002a; Qaim and de Janvry 2003). Technology fees were imposed, and seed was sold at $103 per hectare by a sole supplier. The authors point out that this price is equivalent to a technology premium of $78, approximately the same as what U.S. farmers have to pay for Bt cotton. In addition, while Argentine seed law allows farmers to reproduce their cotton seed for one season before buying new, certified material, the seed supplier prohibited the use of farm-saved seed (Trigo et al. 2002a; Qaim and de Janvry 2003).

The methods applied in the Argentina case are exemplary in their disciplinary excellence. Qaim and de Janvry (2003) applied an approach that combined stated and revealed preferences to estimate farmers' willingness to pay (WTP) for Bt seed. After constructing seed demand functions for farmers and profit functions for the supplier, they were able to demonstrate that both farmers and the monopoly supplier would have been better off at a lower seed price. In one of the most comprehensive approaches applied in the literature, drawing from personal interviews conducted with 299 cotton growers, the authors (Qaim, Cap, and de Janvry 2003; Qaim and de Janvry 2005) estimated an insecticide use function, an instrumental variable model with insecticide use and a production function, and a production model with damage abatement. They compared estimated parameters across larger and smaller farms and, within adopting farms, between Bt and non-Bt plots. By evaluating conditions on Bt and non-Bt plots operated by the same farmer, selection and placement bias are effectively controlled. The authors found that large family businesses benefited from Bt cotton primarily through reduced pesticide use, since pesticide use is correlated with farm size. They predicted that smallholders, who were not then using the technology, could have attained higher gross benefits per hectare because of substantial yield advantages (up to 42 percent). The authors also incorporated a physiological model of the Bt cotton-test system calibrated with entomological data, in order to draw inferences about the size of the refuge areas needed to ensure that farm benefits are sustained.

In terms of findings, the case of Argentina has limited applicability to other developing economies. As compared to the smallholder farmers of China, India, and South Africa, Bt cotton adopters in the study by Qaim and de Janvry (2003) farmed an average of 730 hectares, and none had a landholding of less than 90 hectares. Typically they ran the farm as a family business and employed one or more permanent workers (Qaim and de Janvry 2003).

Other Crops and Traits

HT Soybeans

Despite the fact that RR soybeans are the predominant genetically modified (GM) crop worldwide, and numerous analyses of international trade include RR soybeans (Chapter 6), there are few peer-reviewed studies that analyze their social and economic impact in either developed or developing countries (Table 3.5). Most studies of impacts on farmers focus on the United States and Argentina. Other preliminary studies have considered experiences in soybean-producing countries like Brazil and Romania.

Qaim and Traxler (2005) provided one of the first studies on the economic impacts of RR soybeans in Argentina. Based on a survey of 59 farmers in three soybean-growing regions, which was conducted in 2001, the authors used partial budgeting methods to analyze the impact of RR soybeans on yields, production costs, and gross margins. By considering farmers' experiences with and without the beans over a three-year period, the authors established that there were no significant differences in soybean yields. While herbicide applications in RR cultivation were higher, herbicide costs per hectare were significantly lower for RR as compared to conventional soybeans. Similar to experiences in the United States, they found that while glyphosate applications had increased, the number of applications from other herbicide families had decreased. One of the main reasons for higher glyphosate applications was the increase in no-till farming practices in Argentina. Seed costs were higher for RR soybeans, but seed price differentials were considerably lower compared to the United States. This is attributable to the fact that Monsanto does not have patent rights on RR soybeans in Argentina. Considering gross margins, Qaim and Traxler (2005) found that on average RR soybeans had an income advantage of about $23 per hectare. Margins were not biased toward large-scale farmers.

Penna and Lema (2003) employed partial budgeting to establish the impact of RR soybeans in Argentina. These authors used data from the Institute of Agricultural Technology in Argentina over the period 1998–2000. By comparing conventional and RR soybean cultivation under till and no-till conditions, they found that a combination of RR soybeans and no-till farming led to the highest gross margin in all three years. Similar to conditions in the United States, they showed that yield differences between RR soybeans and conventional soybeans are on average small. To account for uncertainties in the calculation of gross margins, Penna and Lema (2003) used Monte Carlo techniques to simulate gross margins under different cultivation scenarios and to establish cumulative distribution functions for these scenarios. Parameters for yields and prices used in the simulations were obtained through expert interviews. The authors demonstrated that the gross margin for RR soybeans cultivated in the major growing season under no-till practices was stochastically dominant over that for all other combinations of cultivation practices. No-till practices may also entail environmental benefits, because they decrease soil erosion.

IR Maize

Maize is grown largely as a commercial crop in the Philippines (Table 3.5). Cabanilla (2004) has estimated the potential impact of Bt maize on farms in the Philippines with a mixed-integer programming procedure, culling data from representative farms. Yorobe and Quicoy (2006) estimated the partial productivity impact of Bt maize in the Philippines by fitting a Cobb-Douglas production function to sample data from 107 Bt and 363 non-Bt growers in four provinces of the country (Bukidnon, Camarines Sur, Isabela, and South Cotabato). Data were collected during the wet and dry seasons of 2003–04. The researchers controlled for agroclimatic factors by randomly selecting non-Bt farms adjacent to adopters. Recognizing that adopters self-select into the adoption group, they applied a two-stage Heckman procedure. In the first stage they predicted adoption with a set of explanatory factors. In the second they estimated the impact on net returns with an equation that included farm financial variables as well as the predicted probabilities from the first stage and the inverse Mills ratio. The authors concluded that per-unit yields and incomes were higher among Bt growers and insecticide expenditures were lower. Major determinants of adoption were risk perceptions, education, training, and use of hired labor. "Increasing the probability of adoption by 10% increased net farm income by 4.1%," an adoption elasticity that is "higher than those observed in developed countries" (Yorobe and Quicoy 2006, 266).

South Africa is the first developing economy to release a genetically engineered food crop. Gouse et al. (2005b, 2006) present the first few years of evidence about Bt (white) maize adoption and impact among large- and small-scale farmers in South Africa, beginning in 1998. Using only farm survey analyses, they find that yields are higher for both groups and pesticide applications are reduced, particularly for large commercial farmers. In the 2006 article they emphasize the consumption characteristics of white maize, noting that the highest-valued benefits from yield advantages were earned by farmers who grind maize for home consumption. Furthermore they remark that in the last season, the fourth consecutive season with scant rainfall, Bt maize growers and growers of non-Bt hybrids produced similar yields. This last finding reinforces the point, also made by other researchers, that the advantages of a Bt variety will depend on the extent of pest pressure.

Table 3.5 Study descriptors: Impacts of other transgenic crops on farms

Crop	Authors	Year	Data type	Sample size	Methods
Soybeans					
Argentina	Muñoz, R.	2004	—	—	Farm survey analysis
	Penna, J. A, and D. Lema	2003	—	—	Stochastic partial budget
	Qaim, M., and G. Traxler	2005	Pilot survey, key informant	59 farms; 118 plots	Farm survey analysis, economic surplus, large open economy, three regions, institutional analysis
Brazil	Contini, E., M. J .A. Sampaio, and A. F. D. Avila	2005	Public data	n.a.	Simulation, cost-benefit analysis
Maize, cotton					
South Africa	Kirsten, J., and M. Gouse	2003	Key informant, field trials	n.a. sample	Review, findings, institutional analysis
Maize					
Chile	Díaz Osorio, J., R. Herrera, J. Valderrama, and J. L. Llanos Ascencio	2004	Farmer association, company data	10	Partial budget analysis, sensitivity, partial equilibrium
Mexico	Birol, E., E. R. Villalba, and M. Smale	2007	Statistical survey	420	Farm survey analysis, stated preferences, choice experiment, latent class model, farm survey analysis
Philippines	Cabanilla, L. S.	2004	Published data	n.a.	Mixed integer programming
	Yorobe, J. M., and C. B. Quicoy	2006	Statistical survey	470	Farm survey analysis, Cobb-Douglas production function, Heckman model, selectivity bias, producer surplus
South Africa	Gouse, M., C. E. Pray, D. Schimmelpfennig, and J. Kirsten	2006	Statistical survey	368; 104; 196	Farm survey analysis
	Gouse, M., C. E. Pray, J. Kirsten, and D. Schimmelpfennig	2005b	Case study survey	33	Farm survey analysis, key informant

Eggplant					
India	Kolady, D. E., and W. Lesser	2007	Statistical survey, company-managed field trials	290	Farm survey analysis, damage abatement production function, value of marginal product, public-private partnership, open-pollinated varieties hybrid comparison
	Krishna, V. V., and M. Qaim	2007	Statistical survey, field trials	360	Farm survey analysis, trial data analysis, double-bounded dichotomous choice, choice experiment, public-private partnership, open-pollinated varieties hybrid comparison
	Kolady, D. E., and W. Lesser	2006	Statistical survey	290	Farm survey analysis, random utility model, bivariate probit, contingent valuation, public-private partnership, open-pollinated varieties hybrid comparison
Other crops					
China	Huang, J., R. Hu, S. Rozelle, and C. E. Pray	2005a	Statistical survey of on-farm trials	347	On-farm trial survey analysis, pesticide use model, yield model
Kenya	Qaim, M.	2001	Pilot survey, key informant	47	Farm survey analysis, economic surplus, closed economy, cost-benefit analysis, sensitivity
Indonesia, Philippines	Adiyoga, W., M. Ameriana, S. R. Francisco, C. B. C. Mamaril, J. M. Yorobe Jr., and G. W. Norton	2006	Trial data, farmer survey, key informant, published data	n.a.	Economic surplus model
Uganda	Edmeades, S., and M. Smale	2006	Statistical survey, key informant	540	Farm survey analysis, ex ante adoption model, simulation

Note: n.a., not applicable; — indicates not specified.

Conclusions

During the first decade of their use by smallholder farmers in developing economies, peer-reviewed research has indicated that, on average, transgenic crops—and in particular Bt cotton—provide economic advantages for adopting farmers. The close of the first decade marks a convenient juncture to review the methodological limitations associated with the first generation of studies, most of which are recognized by authors themselves. These limitations have implications for the findings, and thus for policy formulation. They should also be addressed in the next generation of studies.

General Caveats

Several general caveats are useful to remember when interpreting the findings in this literature. First, the magnitude of the economic advantages associated with these crop-trait combinations varies widely according to the nature of the cropping season and the geographical location of the study. As is documented in a number of studies, and would be the case for any new technology, averages can disguise the obvious fact that not all farmers profit from adoption. The extent of gains depends on the other varieties available to the farm, farming practices, the severity of pest infestations, and seed prices.[4] Numerous analyses highlight the heterogeneity of farms, farmers, and markets. The latest studies during the decade have begun to explore this heterogeneity more systematically.

Second, the length of the period over which adoption and impact are observed can dramatically shape results. The effects of seed technical change in farming communities are difficult to establish both because of the direction of causality and because, with the passage of time, indirect effects occur that are not visible in earlier years.[5] Many success stories in individual communities are short-lived or episodic; on the other hand, the full contribution of investments in research on an agricultural economy may not be evident until several decades have passed (see the industry studies in Chapter 4).

---

[4]Periodic reviews of the impacts on U.S. farmers, using larger datasets over a longer time period, with analytical methods that are well suited to the agricultural economy, reach the same conclusion (Klotz-Ingram, Jans, and Fernandez-Cornejo 1999; Shoemaker et al. 2001; Fernandez-Cornejo and Caswell 2006).

[5]For example, a first round of studies on the effects of the Green Revolution in Asia found increasing inequality of assets and income distributions (Griffin 1974). Second-generation studies concluded that, at least in the more favorable production areas, absolute poverty declined when food price effects and indirect linkages to the rural nonfarm economy were taken into account (Pinstrup-Andersen 1979; Lipton and Longhurst 1989; Mellor and Johnston 1984; Hazell and Ramaswamy 1991). One general conclusion from the experiences of the Green Revolution is that it is most often the underlying social structure that predetermines much of the social impact of technology adoption.

It is also true that the impacts observed depend on the particular point along the adoption path that is analyzed. In the initial years of adoption, researchers have focused on examining "first-round" impacts. Establishing that transgenic crops are profitable, that yields are advantageous, and that adoption of IR or HT crops has reduced the use and costs of inputs (labor, herbicides, and pesticides) is of paramount importance. Measurement of impacts on poverty, inequality, health, and the environment gains in importance and in feasibility as adoption continues. So far these issues have been addressed in fairly simple ways, with indicators rather than within an economic theory or framework. For example, not a single study reviewed examined differential impacts on men and women.

The range in transgenic crops studied and in the study contexts should also be borne in mind. Studies of Bt cotton, which has unique economic and agronomic properties, dominate the literature, along with a few country case studies. We should be careful not to generalize from these experiences to other crop-trait combinations and contexts.

Similarly there are relatively few *different* authors publishing case studies in peer-reviewed, international journals. The fact that there are fewer peers suggests that there is a narrower consensus of opinion. There is also a wide range in the quality of publications. In some cases the political sensitivity of the issues surrounding transgenic crops makes it difficult to implement research in particular countries and farming communities. This operational difficulty compounds the more specific forms of selection bias, mentioned in the next section.

Specific Limitations

Selection bias. Most of the first-generation studies exhibit potential bias associated with study placement, selection of farmers through a company extension program, and/or self-selection of certain types of farmers into the adopting group. Authors have sought to address these biases in various ways, and with greater depth in later publications. Methods for minimizing these biases are thoroughly explored in the broader economics literature, and these should be brought to bear explicitly on this topic.

Even if these biases have been mitigated through study design, identifying the appropriate counterfactual for variety and practices has not been straightforward. As the decade progressed, there was ample recognition that the variety used for comparisons, and the range of inputs monitored by the researchers, influenced the size of the estimated advantages associated with transgenic crops. Establishing quasi-experimental conditions to compare isogenic lines with transgenic varieties grown by the same farmers in multiple locations may be the preferred method for estimating yield advantages with precision, but this approach removes the farmer from an actual decisionmaking context. In large-scale surveys, variables to control for the fixed effects of germplasm may enable researchers to improve impact estimates.

Measurement bias. When sample sizes are small, as is the case in a number of studies, errors associated with sample design can be great. By contrast, measurement errors are often proportionately greater in larger-scale surveys. The use of recall methods to measure input use and yields has often been criticized; many of the studies appear to rely on this technique, despite the fact that these are the key parameters of interest for measuring the impacts of IR or HT crops. This problem may be exacerbated by situations in which farmers do not read labels or mix chemicals from different containers. Several authors have insisted on the importance of monitoring practices daily, in order to develop a more realistic picture of the full range of biotic pressures (several bollworms, sucking insects, other fungal diseases). Several studies note the importance of research protocol for measuring agronomic practices, and some authors have undertaken careful (and expensive) monitoring routines. When researchers have sampled plant tissue, they find that toxin expression varies significantly not only by season but also by cultivar and plant part—a crucial methodological finding. Such techniques, as well as objective yield measurements, are expensive to implement relative to those that rely on recall, but they must be carefully considered. These tasks will only become more daunting as the prospect of secondary pest evolution emerges with continuous adoption.

Estimation bias. Partial budgets are deceptively simple. In fact considerable care must be used to construct them. In a number of the studies examined here, only gross margins are reported. Gross margins include the costs of intermediate inputs but ignore the use of labor and land. Net margins include these costs. The way in which use of transgenic crops affects budget categories depends on the particular crop-trait combination and cannot be generalized.

By definition partial budgets treat only one farm activity at a time. Even where farmers are fully commercialized, the net impact of adoption on whole-farm production and resource use cannot be deduced from a partial budget. Cross-activity impacts have not been systematically investigated.

Even when whole-farm production is considered, when farmers are not fully commercialized, and when they operate in situations with market imperfections, the input and output prices that influence their decisions are endogenously determined and household-specific. This consideration is likely to be of greater significance for subsistence food crops. So far a household model has been applied only to the case of transgenic cooking banana in the East African highlands. Finally, risk and uncertainty have been considered explicitly in only a few of the studies, in which stochastic budgets were developed and Monte Carlo simulations were conducted. Given the consistent evidence of outcome variability, examining the statistical distributions of impact variables (yields, costs, profits), in addition to average impacts, seems fundamental for future work. In general, risk theoretic approaches are largely absent in this first generation of studies.

Endogeneity problems haunt applied economists, and they are especially evident in the case of pesticide use if farmers apply pesticides in response to pest pressure. The decisions to grow an IR variety, as well as to apply pesticides or herbicides, are likely to be affected by the same variables that influence the yield of the transgenic crop, net returns, or other outcome variables. If sources of endogeneity are not tested and treated in the econometric specification, regression coefficients will be biased. This problem has often been recognized, but not always addressed, in the literature assessed here. The use of production functions with damage abatement in later analyses is a major improvement, since these recognize explicitly that pesticides abate damage rather than enhance productivity. Modeling is made more difficult when the yield-saving effects of the transgene cannot be distinguished from the yield-enhancing effects of the host germplasm.

Even though multiyear surveys were conducted in a number of country cases, attempts to model the dynamics of adoption and learning, and to control for farmer- and year-specific effects through the use of panel data, are apparent in only a few studies. Year-to-year variability of impacts, in addition to cross-sectional variability, is a salient finding, in particular for transgenic crops with insect resistance.

A final comment concerns whether adoption is supply- or demand-driven, a distinction that is crucial for designing policies to support the adoption of proven technologies. Findings clearly point to the hypothesis that arrangements for supplying seed and purchasing the product, such as the extent of vertical coordination and competition, affect adoption and farm impacts. Linked to sector analysis and the evaluation of market channels, this aspect of adoption clearly needs greater emphasis in studies conducted on farms.

Chapter 4

Impacts on Consumers

Approaches

Two main bodies of literature address the influence of transgenic crops on consumer behavior, neither of which assesses impacts per se. Consumers represent the market demand of end users for farm products, which, combined with supply, determines price. Without consumer demand for biotech products, farmers will not be able to sell their crops on the market even if they adopt them. For this reason, consumer attitudes and perceptions play a fundamental role in the global debate concerning transgenic crops. The information consumers are able to assimilate critically affects their preferences. Conventional wisdom, backed by empirical research, is that consumers in developing economies are less well informed about biotechnology than their counterparts in industrialized countries. The relative cost of obtaining information is one explanation for this discrepancy, but government policies toward public awareness and education undoubtedly play a role. Over time, as more information accumulates and consumers gain knowledge, preferences—and hence market demand—will shift.

The first body of articles consists of surveys designed to elicit the attitudes of consumers toward products made with transgenic crops (here we refer to these as biotech foods). Surveys record whether or not consumers are concerned about transgenic crops and the nature of their concern. These data, which are generally descriptive in nature, provide essential baseline information for policymakers.

In the second set of articles, authors exploit recent advances in stated-preference methods to estimate consumers' WTP for products that are free of these ingredients. These articles are intended to provide decisionmakers with estimates of the price premium that would be necessary to market a transgenic product successfully. Estimates of WTP are useful in marketing research, and they are needed in order to assess the welfare implications of labeling policies. The transgenic product is a close substitute for the nontransgenic product, resembling it in all attributes except that at least one ingredient is derived from transgenic raw materials. Most of the methods applied in the second body of literature elicit hypothetical choices using carefully constructed menus of options, or choice sets, presented to consumers. Researchers recognize that there is often a difference between what people state they will do and what they actually do. Some of the most recent advances in these methods involve combining stated-preference methods, which are based on hypothetical situations, and revealed-preference methods, which record actual situations.

The use of stated-preference models in environmental economics and marketing research also continues to advance. For example, Kontoleon (2003) found that the latent segmentation model is superior statistically to other methods, including (1) multinomial logit with interacted individual characteristics, (2) random parameter logit, (3) covariance heterogeneity models, and (4) latent class models. Tapping choice experiment data collected from a sample of consumers, the latent segmentation model enables the researcher to segment the population of consumers into distinguishable groups and at the same time explain the choices made by each population segment. Understanding the heterogeneity of consumer perceptions supports the design of educational programs and the formulation of effective marketing strategies.

Aside from the two groups of literature on consumer behavior, four studies examine the potential impact of biofortified crops on public health in Asia. A list of the publications by group is provided in Table 4.1. Including all articles identified in the search, only 28 were conducted in countries with nonindustrialized agriculture. By contrast, at least 61 articles have been published on the same topic in industrialized agricultural economies. As was the case for the studies summarized in Chapter 2, China is by far the most heavily represented country. India, Korea, the Philippines, and Taiwan also represent Asia. No studies conducted in Africa met the selection criteria (Chapter 1). Only three studies are reported for Latin America: Mucci, Hough, and Ziliani (2004) for Argentina, Pachico and Wolf (2004) for Colombia, and Aguilar and Kohlmann (2006) for Costa Rica.

Consumer Perceptions

Zhong et al. (2002), Zhou and Tian (2003), and Zhang (2005) surveyed the perceptions of consumers in Beijing, Nanjing, Shanghai, and Tianjin. In the survey led by Zhang, approximately 300 consumers were asked how much extra they were willing to pay for vegetables that were (1) pollution-free, (2) green, (3) organic, and (4) biotech. Of these categories, consumers were least willing to pay more for biotech food. An experimental design was not used to estimate WTP. The authors

Table 4.1 Study descriptors: Consumer acceptance and willingness to pay for biotech food

Authors	Year	Country (city or region)	Biotech product	Sample size
A. Consumer perceptions				
Aguilar, F. X., and B. Kohlmann	2006	Costa Rica	Banana	101; 27
Govindasamy, R., B. Onyango, W. Hallman, H.-M. Jang, and V. Puduri	2004	Korea	General	903
Ho, P., and E. B. Vermeer	2004	China (Beijing, Shijiazhuang)	General	1,000
Ho, P., E. B. Vermeer, and J. H. Zhao	2006	China (Beijing, Shijiazhuang)	General	1,000
Hu, W., and K. Chen	2004	China (Beijing)	Vegetable oil	671
Mucci, A., G. Hough, and C. Ziliani.	2004	Argentina	General	250
Onyango, B., R. Govindasamy, W. Hallman, H.-M. Jang, and V. S Puduri	2006	Korea	General	570
Pachico, D., and M. M. Wolf	2004	Colombia (Cali)	General	150
Zhang, X.	2005	China (Tianjin)	Vegetables	300
Zhong, F., M. A. Marchant, Y. Ding, and K. Lu	2002	China (Nanjing)	General	480
Zhou, F., and W. Tian	2003	China (Beijing)	General	
B. Consumer willingness to pay				
Anand, A., R. C. Mittelhammer, and J. J. McCluskey	2007	India	Wheat	600
Chern, W. S., K. Rickertsen, N. Tsuboi, and T.-T. Fu	2002	Taiwan (Taipei)	Vegetable oil	213
Curtis, K. R., and K. Moeltner	2006	China	Rice, Soy oil	598
Deodhar, S. Y., S. Ganesh, and W. S. Chern	2007	India	General	602; 102
Hu, W. Y.	2006	China (Beijing)	Vegetable Oil	570
Hu, W. Y., F. Zhong, and Y. Ding	2006	China (Nanjing)	Soy oil	449
Huang, J., H. Qiu, J. Bai, and C. E. Pray	2006	China (Eastern)	General	1,005; 666
Jan, M., Fu, T.-T., and C. L. Huang	2007	Taiwan	Tofu	940
Li, Q., K. R. Curtis, J. J. McCluskey, and T. I. Wahl	2002	China (Beijing)	Rice	599
Lin, W., A. Somwaru, J. Huang, and J. Bai	2006	China (eastern coast)	Rice	1,100
Lusk, J. L., M. Jamal, L. Kurlander, M. Roucan, L.Taulman	2005			n.a.
McCluskey, J. J., K. M. Grimsrud and T. I. Wahl	2006	China (Beijing)	Rice, soy oil	599
C. Impact on nutrition				
Bouis, H.	2002	India, Bangladesh	Rice, wheat	
Dawe, D., R. Robertson, L. Unnevehr	2002	Philippines (Cebu)	Rice	1,839
Stein, A. J., P. Nestel, J. V. Meenakshi, M. Qaim, H. P. S. Sachdev, Z. A. Bhutta	2007	India	Rice and wheat	
Zimmermann, R., and M. Qaim	2004	Philippines	Golden rice	

Note: n.a., not applicable.

found that young, highly educated respondents were more likely to be willing to buy biotech food, but because they perceived that it would be of higher quality. Zhou and Tian (2003) report that consumers in Beijing had some general knowledge of biotech food, and many misunderstandings, but attitudes were optimistic, especially if they believed that product quality would be enhanced, pesticide use reduced, and the environment protected. Zhong et al. (2002) used telephone interviews of over 500 households in the Nanjing urban area, supplemented by an analysis of Chinese media reports. Their results indicated that the majority of Chinese consumers had little knowledge of biotech foods, that almost all thought they should be labeled, and that media attention to the subject has increased since the late 1990s in both frequency and the number of articles transmitting negative messages. After being given general information, 40 percent stated that they would buy, 17 percent would not, 8 percent would follow the majority, and the remainder did not know. When provided with more specific information, consumers cared more about nutritional enhancements or medical functions, which affect them directly, than pest resistance. Older respondents tended to accept biotech foods more than younger respondents, as did men compared to women.

Hu and Chen (2004) surveyed consumers' intentions to purchase biotech vegetable oil in Beijing, interviewing 671 respondents in 2002–03 in farmers' markets and supermarkets. They noted that the demographic statistics of the sample are comparable to those reported in the 2001 *Beijing Statistical Yearbook.* In a survey with several stages and a randomized ordering of questions, they sought to answer questions concerning how consumers' attitudes toward biotechnology, their knowledge, and different types of information affect their purchase intentions. They estimated a multinomial logit model with three options (purchase, do not purchase, "don't know"). Findings indicated that more than 67 percent of respondents were concerned about biotech foods, and about 20 percent believed that biotech foods are harmful to health. The regression model suggested that the more consumers trusted the national food safety system, the more likely they would be to purchase biotech oil. The presence of favorable information had a significantly positive effect on intentions to purchase, as did older age, more knowledge about biotech foods, and more education in general. The authors note that they estimated only the effects of favorable information on consumers' intentions, and that the influence of unfavorable information constitutes another important topic.

Govindasamy et al. (2004) applied two ordered probit models based on random utility theory to data collected by Gallup Korea through personal interviews with 1,054 adults from 20 to 59 years in age. They tested the effects of social, economic, and demographic variables on consumer approval of genetic modification involving both plant and animal genes. Regression results indicate that consumers with above-average knowledge of specific outcomes of genetic modification were more likely than those with inaccurate or no knowledge to approve of its use in either plant or animal products. Furthermore consumers who trust institutions, farmers, and the media also have more favorable attitudes. Thus the authors recommend campaigns to educate consumers. Those who felt labeling would be necessary were also less likely to approve of genetic modification. Like other authors, Govindasamy et al. (2004) found that women were less likely to approve of the technology. As can be expected, differences in approval scores among consumers are associated with residential area, income, education, and political affiliation.

Ho and Vermeer (2004) surveyed 1,000 residents in four supermarkets of Beijing and Shijiazhuang concerning their awareness and acceptance of biotech foods during 2003. Although 71 percent reported that they were aware of biotech foods, only 32 percent could mention a biotech crop (most frequently soybeans, soy, or vegetable oil). Despite China's importance as a producer of Bt cotton, very few respondents cited this crop. Less than one-fifth of the sample could mention reasons why farmers grow biotech crops, and over 80 percent "had no inkling about genes and could not correctly answer whether the statement 'nontransgenic soybeans do not contain genes, but transgenic soybeans do' was true or false" (Ho and Vermeer 2004, 169).

In this article and a 2006 article by Ho, Vermeer, and Zhao, the authors express concern that consumer resistance may eventually develop in China. They point out the "malleability of the Chinese consumer in a context of limited understanding and inadequate access to information" (Ho, Vermeer, and Zhao 2006, 227). The authors review the political economy of food regulations and biotechnology in China, along with recent experiences with unsafe food. Based on their survey of 1,000 respondents, they find that while the initial estimated percentage of consumers who were willing to purchase biotech food in China was much higher than in other countries where it is marketed, this percentage dropped substantially when respondents were provided with additional, neutrally worded background information. Responses to a range of questions also demonstrated that when consumers reported an awareness of genetic modification, they did not necessarily understand its meaning.

Onyango et al. (2006) explored the heterogeneity of consumer attitudes to biotech foods in South Korea using factor and cluster analysis. The questionnaire administered, through personal interviews by Gallup South Korea, had been implemented earlier in the United States, and was modified for cultural differences. Six factors accounted for 61 percent of the variation in responses: (1) environmental, taste, and price benefits; (2) the importance of "naturalness" in foods; (3) convenience and "comfort" in food; (4) perceived risks of various types; (5) open-mindedness about biotechnology; and (6) open-mindedness about new foods. They identify four consumer segments, including "consumers who are ardent supporters of the naturalness attribute of food, consumers who are apprehensive about biotechnology, those who are food adventurous, and consumers seeking more information on biotechnology" (Onyango et al. 2006, 76).

Studies implemented in Latin America are few. In 2001 Pachico and Wolf (2004) conducted a preliminary investigation of the attitudes of 150 consumers in Cali, Colombia, regarding biotech food. Food purchasers had a positive view of science and technology and were confident that the government would assure food safety. Data suggest that 38 percent of respondents are not always able to provide the amount of food they would like for their families, and 55 percent note that price is the most important factor in making purchase decisions. These results underscore the fact that consumer choices, whether in favor of or against biotech products, are strongly constrained by low incomes in many developing countries. Only a minority of respondents in the Pachico and Wolf study explicitly stated their unwillingness to buy biotech food, even though three-quarters of them perceived it to be potentially risky. The authors remark that familiarity with biotech food was still very low in their sample at the time of the study.

Mucci, Hough, and Ziliani (2004) interviewed 250 consumers in Buenos Aires regarding their intent to purchase biotech foods and related perceptions. To overcome difficulties encountered when requesting personal interviews, they used a "drop-off" method. Respondents were personally contacted to request a survey, the survey was left for them to complete, and the enumerator returned to retrieve the findings and offer a gift as compensation for the respondents' time. Three-quarters of respondents stated that they knew about biotech food, but perceptions were generally negative. Intention to purchase was influenced positively by the promise of nutritional benefits, trust in the brand, younger age, fewer years of education, and lack of knowledge about biotech products.

Aguilar and Kohlmann (2006) asked consumers about their willingness to consume transgenic bananas (as compared to their WTP) and asked producers about their willingness to adopt them in Costa Rica. Both questions were hypothetical. They then identified the factors that explained the choices made by respondents with a probit model and a mixed linear regression. Younger, wealthier, more highly educated consumers were likely to state that they would consume transgenic bananas. All farm managers surveyed expressed interest in transgenic banana varieties owing to the high cost of pest management in this crop. Their estimated WTP for the cost of such varieties, which ranged from US$500 to US$999 per hectare, was related only to these costs and not to farm or farmer characteristics.

Consumer Willingness to Pay

As a useful point of reference, Lusk et al. (2005) conducted a meta-analysis of 25 studies on consumer demand for transgenic food, all carried out in industrialized agricultural economies. The authors concluded that (1) consumer characteristics, (2) the method used by the researcher, and (3) the food studied explained 89 percent of the variation in the estimated WTP for food that is free of transgenic ingredients. They proposed a meta-model as an economical means of generating rapid estimates for policymakers with reasonable accuracy.

Mandatory labeling of biotech foods makes it important to know the premium consumers are willing to pay for non-biotech foods when they prefer these to biotech foods. So far the authors who have sought to quantify WTP for non-biotech foods in developing economies are few. Chern et al. (2002) contrast consumer acceptance and WTP for biotech vegetable oil and salmon in Japan, Norway, Taiwan, and the United States. The authors applied a choice experiment methodology to data collected from samples of 100–200 university students in each country. They estimated that student consumers in Taiwan were willing to pay a premium of only 17–21 percent to avoid purchasing these biotech products, as compared to 33–40 percent in Japan, 50–62 percent in the United States, and 55–69 percent in Norway.

Curtis and Moeltner (2006) compared the propensity of consumers to purchase biotech food in China and Romania by estimating a consumer value function for biotech substitutes, including rice and soybean oil (for China) and potatoes and sunflower oil (for Romania). Consumers were asked about basic demographic factors and their attitudes toward science; they were also asked dichotomous-choice, contingent-valuation questions regarding their willingness to accept or their WTP for biotech food products. Bivariate probit models were estimated. According to the authors, perhaps their most "striking result" is that while only about 8 percent of Chinese consumers consider biotech products to be "high risk," "the picture for Romania is almost diametrically reversed, with only 11% of shoppers perceiving no risks" (Curtis and Moeltner 2006, 298). One of their conclusions is that risk perceptions are the major drivers of willingness to participate in biotech food markets, while standard demographic variables are poor indicators. They suggest that in the Chinese case, a policy aimed at growing biotech crops along with an economically feasible segregation system for biotech and non-biotech products would prove possible, given that U.S. and Chinese consumers accept biotech products while consumers in the European Union (EU) demand non-biotech products.

Li et al. (2002) interviewed 599 consumers in four locations (a supermarket, two outdoor markets, and a shopping area) of Beijing using the dichotomous-choice, contingent-valuation method, considering soybean oil and product-enhancing rice products. They estimated a double-bounded logit model, finding a positive mean WTP for biotech foods, in contrast to findings from other countries. Respondents generally had a favorable view of biotech rice and soybean oil, which are staples in the Chinese diet. Younger respondents had a higher WTP. The authors state that consumer attitudes are influenced by "positive media coverage, which is controlled by the government" (2002, 151). They conclude that Chinese citizens are unlikely to change their attitudes owing to limited media coverage of the debates occurring elsewhere.

Building on earlier studies of consumer perceptions, Huang et al. (2006) sampled a subset of respondents to the nationwide Urban Household Income and Expenditure Survey, conducting personal interviews with over 1,000 households in 11 cities in northern and eastern China. Using a survey instrument that had previously been applied in Canada, the EU, Korea, and the United States, as well as careful interview approaches, they achieved a high response rate. First they asked whether the respondent would be willing to buy a selection of hypothetical biotech foods if prices were the same for biotech and non-biotech foods. If the answer was no, the respondent was asked if he or she would be willing to buy the biotech food at a 10 percent discount. If the respondent also answered no to this question, he or she was asked at what discount the biotech food would be acceptable. The authors found that despite awareness of transgenic foods, consumer knowledge was limited; however, Chinese consumers demonstrated a greater acceptance of and WTP for transgenic foods than is evident in other countries. Nonetheless 20 percent said they would never buy biotech food, whatever the price.

Lin et al. (2006) analyzed survey data from 1,100 consumers in cities of various sizes, including Beijing and Shanghai. Applying a semi-double-bounded model of dichotomous choice (contingent valuation), they considered eight different biotech foods, with a focus on soybean oil and IR rice. They estimated that 60 percent of respondents were willing to purchase biotech foods without any price discounts, although about 20 percent would accept only non-biotech foods. The mean WTP for non-biotech foods averaged 23–53 percent for this latter group and 42–74 percent for non-biotech rice. With respect to methods, they comment that WTP could potentially be more accurately estimated if consumer awareness were treated as an endogenous variable, since many of the variables that affect attitudes also affect awareness.

Jan, Fu, and Huang (2007) conducted a conjoint analysis to examine the relative importance of product attributes in influencing consumer preferences for tofu, drawing their data from personal interviews in Taiwan, of which 940 were usable. Regression analysis with a binary-choice, random-parameter logit model demonstrated that the biotech attribute was significant but least important after brand name and price. Cluster analysis also revealed three distinct market segments for tofu, of which the largest population segment had a lack of concern. The biotech attribute was more important for other smaller population segments that were either for or against purchasing biotech tofu. Women, in particular, were less willing to purchase GM crops. The authors concluded that many Taiwanese know little about issues related to biotechnology and recommended that greater attention be paid by policymakers to educational campaigns.

Hu, Zhong, and Ding (2006) combined double-bounded and payment card approaches to contingent valuation in testing the hypothesis that consumer WTP for biotech soybean oil in Nanjing changes with new information, using a sample of 523 respondents. The payment card approach corrects for bias associated with lower extreme values of WTP. Since the respondent can indicate the minimum price decrease necessary before he or she would consider buying the biotech product, a broader range of values, including true zeros, is identified. Estimation results illustrated that media reports of "real-life cases" have a significant impact on consumer WTP. Positive reports had positive effects, but these were not nearly so large as the negative effects of negative reports.

Based on the Beijing sample interviewed in 2002–03 (reported previously), Hu (2006) estimated three models in order to treat the endogeneity of purchase and payment decisions, akin to a two-stage, sample-selection approach. The first is the conventional binary-choice model that does not separate the probability of purchase from the amount the consumer is willing to pay and does not account for zero WTP. The second model includes two equations. The first specifies demographic and attitudinal variables to explain the probability that a consumer chooses to purchase the product, and the second includes a constant and the payment card variable to explain payment amounts. The third model recognizes the correlations between the two steps: whether to pay a positive amount for the biotech oil, and, if the answer is positive, how much. The same variables are used in the two stages except for product price. Hu calls this approach the spike WTP model with covariates. Though he finds the statistical results to be similar between the spike model and the conventional binary-choice model—suggesting that there is no bias created by ignoring the participation decision—he favors the spike model because of its treatment of nonparticipation probabilities.

Following a three-year survey of consumers in Canada, Japan, Norway, and the United States using a dichotomous-choice, contingent-valuation approach, Mc-Cluskey, Grimsrud, and Wahl (2006) estimated an ordered multinomial choice model and compared results among the countries. Products considered were noodles and bread made with biotech wheat, and soybean oil and tofu made with biotech soybeans. The three categories considered were consumers who were willing to buy biotech food without a discount, those willing to buy it at the offered discount, and those unwilling to buy it at a discount. While the results for the United States and Canada are similar and those for Norway differ with respect to the estimated effects of some social and demographic factors, the data from China present an entirely distinct picture. On average, Chinese consumers were willing to pay a premium for biotech foods, as was found in the study by Li et al. (2002). The authors explain these findings by noting the positive attitudes of consumers to biotechnology, their low perception of risk, and their unique cultural and political history which leads them to be "forward-looking" and to view "technological novelties from the rest of the world are often considered as much-needed improvements" (Li et al. 2002, 151).

The market for chapati made from biotech wheat was investigated in India by Anand, Mittelhammer, and McCluskey (2007). Using data collected during 600 personal interviews with shoppers in grocery stores in New Delhi and Patna, they tested the effects on WTP of whether or not information was received and whether the genetic modification was portrayed as producer- or consumer-friendly. The method they used was contingent valuation. The authors found that consumers were willing to pay a slight premium for biotech chapati when no information was provided. When "producer-friendly" information was provided, WTP increased by a small amount, but when negative information about potential health effects was offered, consumers reacted strongly and negatively. This effect was offset somewhat when they were given "consumer-friendly" information.

Deodhar, Ganesh, and Chern (2007) analyzed the likelihood that consumers in Ahmedabad city would be willing to consume biotech cottonseed oil, rice, and chicken, and the extent of the premium they would pay to avoid it. Personal interviews with households were combined with a web-based survey of businesspersons, scientists, and students. The contingent-valuation method they applied was based on five sequential, binary, stated choices. Their findings demonstrated that awareness of biotech food was extremely low among Ahmedabad respondents and greater among Internet users. After receiving more information about the pros and cons of biotech foods, more than 70 percent were willing to consume them at the same price as non-biotech products. Most Indian consumers want mandatory labeling, although nearly a third of them are not willing to pay for it. The stated likelihood of consumption was also sensitive to product price.

Impact on Nutrition

Bouis (2002); Dawe, Robertson, and Unnevehr (2002); Zimmermann and Qaim (2004); and Stein et al. (2007) are the first authors to explore the potential impact of biofortified crops on human health and nutrition.

Dawe, Robertson, and Unnevehr (2002) explored the possible role of vitamin A–enriched rice in a case study conducted in the Cebu region of the Philippines, a poor region with severe nutritional deficiencies. The data were selected in part because they were recent and were accurate regarding the intake of an at-risk population, including mothers who had been selected while still pregnant in

1983–84, with follow-up surveys performed from 1991 to 1999. The authors used actual consumption patterns of rice and white maize grits to estimate the effects of substituting enriched foods. Thus potential impact depended on dietary patterns, and it was predicted to be modest in the region of study because rice was not the main staple. The authors emphasize the role of changing diets in Asia, the fact that consumer attitudes were an important unknown, and other technical issues regarding production costs and the stability of beta-carotene during storage. Placing the work in perspective, they conclude that vitamin A–enriched rice, while likely to deliver substantial amounts of vitamin A under certain scenarios, is "unlikely to meet all requirements and would be an ineffective stand-alone strategy" (Dawe, Robertson, and Unnevehr 2002, 557).

Zimmermann and Qaim (2004) and Stein et al. (2007) express the social burden of nutrient deficiency in terms of disability-adjusted life years (DALYs) lost. Zimmermann and Qaim (2004) investigate vitamin A–enriched Golden Rice in the Philippines, using the data on food intake collected by the Food and Nutrition Research Institute (1993, 1998) and postulating a function that relates vitamin A intake to disease levels. (They note that while a general relationship is widely accepted in the literature, concrete evidence on the exact numerical association is lacking.) They estimate that Golden Rice could generate social benefits of US$16–88 million per year through reducing the incidence of blindness and premature death. Combining these data with a cost-benefit analysis of investments in research and development, they find high internal rates of return.

Stein et al. (2007) draw from a nationally representative survey (a sample size of 119,544!) conducted in India in 2000, applying a dose-response function to project the potential health improvements from zinc intakes. Introducing data on costs of development and dissemination compiled by international agricultural research centers, they estimate that the cost of saving one DALY is US$0.73–7.31—which is cost-effective by World Bank and WHO standards.

Conclusions

The number of studies assessing the role of biotech products in influencing consumer demand in developing economies is small relative to those evaluating farm impacts and limited compared to those published in developed economies—in part because a feature of developed economies is a structurally stronger market channel with greater consumer awareness and highly articulated, differentiated preferences.

As is true in the literature on farm impacts (Chapter 3), we find that China is the most intensively studied case in terms of publication counts—an expression of the nation's leadership in the promotion and diffusion of crop biotechnologies.

Several general threads run through this first decade of publications on consumer behavior. The authors have also cited some specific caveats that should be borne in mind for future research.

General Caveats

The focus on China in the literature is important and informative, given the role of the country in generating new biotech products, but China is unique as a case study. Authors appear to be in general agreement that in the first instance—and in part as a reflection of deliberate government policy—consumer attitudes in China have been shown to be more accepting of biotech products than those in other countries. Most authors ascribe this outcome in part to government policies, and some to cultural and political history, although not all agree on whether the policies have played a positive or negative role.

A finding that cuts across a number of studies, whether conducted in China or elsewhere, is the crucial impact of information on the preferences expressed by consumers. Research has demonstrated that the attitudes of consumers change significantly as they absorb new information, and particularly negative messages. This has two obvious implications. First, as in most studies of consumer behavior, framing of questions appears to be of paramount importance. In addition, given the rapidity of change in this field of science and the quantity of information to which consumers are exposed, estimates of perceptions and WTP may need to be continually updated for the information to be of use in marketing.

There is generally a disconnection between products studied in analyses of consumer preferences versus farmer adoption. One reason is that cotton is the most heavily studied product in the literature on farm impacts, yet it is generally not a food crop; among food products, only biotech maize and soybeans have been approved for release in developing economies during this first decade.[1] One study on bananas in Costa Rica attempted to relate the propensity to purchase and the propensity to adopt. In many low-income countries, the length of the market chain is relatively short and farmers are also consumers of some of their own produce. More efforts to link both sides of the market in one study could be fruitful.

In this regard, relatively few studies have sought to estimate WTP with advanced methods until recently. Later studies present more complex statistical tests of hypotheses. As in the analyses of farm impacts, the issue of sample selection bias—or the relationship between the latent variables that predict a decision to participate and the effect of participation on its outcome—has been recognized. So far, all published studies of impacts on nutrition are ex ante.

---

[1]At least in the Sahelian countries, however, parts of the cotton plant are used as food and fodder.

Specific Limitations

Clearly coverage of products and countries is at present too narrow for generalizations to be drawn. While the food types and traits considered in the studies are few, they are also heterogeneous in the sense that the final product may contain different proportions of ingredients derived from transgenic crops. Studies conducted in low-income nations other than China are gradually emerging, and more are needed before generalizations can be drawn.

As was evident for the studies of farm impacts, research on consumer impacts in developing countries could benefit greatly from methodological advances in the general valuation and consumer choice literature, such as combined revealed- and stated-preference approaches. In their meta-analysis of the literature on industrialized economies, Lusk et al. (2005) were able to demonstrate the fundamental point that estimates of consumers' WTP for the biotech attribute are conditional on the method employed by the researcher. In most cases researchers admit that stated-preference approaches tend to overstate WTP, suggesting that the discounts that must be applied to make biotech food marketable have also been overestimated. On the other hand, revealed-preference approaches are known to suffer from statistical shortcomings and can only be implemented where biotech food products have already been marketed. A combined approach cannot be implemented until biotech food products have been released, but it could dramatically improve the reliability of estimates through data augmentation.

The approach recommended by Lusk et al. (2005) merits attention if it can be employed at lower cost to generate the minimum information needed within acceptable confidence intervals. For that to be feasible, however, additional studies with different methods will need to be conducted. The fact that most published studies have been implemented in industrialized economies does not in itself suggest that the methods are inappropriate for consumers in developing economies. However, large-scale mail, phone, or Internet surveys are cheap compared to the personal interviews that would most likely be needed to study consumer preferences in developing economies. Some of the challenges described in the Buenos Aires and Cali studies illustrate this point.

Chapter 5

Impacts on the Sector

Approaches

Two dozen peer-reviewed publications address the impacts of a transgenic crop on the crop sector, including both growers and consumers of the product, in a developing economy. Nine of these present estimates of farm impacts from sample surveys as direct inputs into the sector analyses (many of these were already mentioned in Chapter 3). One article (Zimmermann and Qaim 2004, discussed in Chapter 4) combines some analysis of effects on consumer nutrition with a consideration of sector impacts. In addition three publications examine regulatory costs and impacts on seed markets. Most of the sector studies discussed in this chapter have been designed to examine impacts ex ante rather than ex post. Thus their purpose has been to provide investors and policymakers with initial estimates of the magnitude of economic benefits and their social distribution.

Across these studies, the primary means of estimating sector impacts has been the economic surplus approach detailed by Alston, Norton, and Pardey (1998). Also termed a partial equilibrium displacement model, the approach considers only the effects of adopting biotech crops in the product market where the change occurs. Effects in other markets, including input markets and markets for substitute or complementary goods, are disregarded. In the standard model, the estimated magnitude and distribution of the economic benefits depend on a number of parameters. These include the price elasticities of supply and demand for the crop, whether the country is a large or small producer (price setter or price taker), whether the country trades the crop internationally (closed or open economy), adoption rates, and the nature and magnitude of the supply shift induced by adoption. The direction and extent of the supply shift are determined by changes in yields and costs induced by the adoption of the technology. Parameter estimates are typically drawn from some combination of sample surveys of farmers, trial data, key informant interviews, and secondary data. The analysis of the sector may be conducted on a regional, national, or global scale.

Over time, in an effort to adapt the economic surplus approach to particular empirical contexts, researchers have proposed numerous modifications of the basic model. For example, Hayami and Herdt (1977) made an adjustment to the basic model to account for subsistence production in a country with nonindustrialized agriculture. Their adjustment partitions the aggregate supply curve into partial supply curves that represent subsistence and commercial growers, with different adoption rates and corresponding impacts on farm income.

One specific feature of GM crops is the capital investment on the part of the innovator required to develop the transgenic product.[1] To ensure that returns on this investment are achieved, private companies have sought to protect their intellectual property. IPRs confer a temporary monopoly in the market. Moschini and Lapan (1997) proposed a theoretical framework to accommodate this feature. Applying this framework with an economic surplus model to the case of Bt cotton in the United States from 1996 to 1999, Falck-Zepeda, Traxler, and Nelson (1999, 2000a, 2000b) laid out a model that has since provided the foundation for other applications in developed and developing economies.

Aside from economic surplus approaches, two other techniques have been employed to examine sector impacts in the literature that met our criteria (see Chapter 1). Huang et al. (2004) applied the GTAP (Global Trade Analysis Project) model to the Chinese national economy. This study is grouped with other trade studies in Chapter 6. Cabanilla, Abdoulaye, and Sanders (2005) developed a linear programming model to estimate the potential impact of Bt cotton in West Africa. This study is grouped with other ex ante analyses in the next section.

A small but growing set of studies addresses the costs, but not the benefits, of regulations (Pray, Ramaswami, and Bengali 2005; Pray et al. 2006; Manalo and Ramon 2007). So far, empirical analyses do not appear to be grounded in a theory of regulatory economics. Only two studies conducted in India have treated issues related to the supply chain for transgenic crops (Kambhampati et al. 2005; Murugkar, Ramaswami, and Shelar 2007).

Descriptors for studies on sector impacts are summarized in Table 5.1.

[1]A "transgenic product" consists of a gene construct, a transformation protocol, and the selection and multiplication of successful insertions, in addition to the plant genetic resources that serve as host for the gene.

Table 5.1 Study descriptors: Industry (sector) impact of transgenic crops

Authors	Year	Crops	Data type	Time	Scale	Country	Methods
A. Impact on crop sector							
Adiyoga, W., M. Ameriana, S. R. Francisco, C. B. C. Mamaril, J. M. Yorobe Jr., and G. W. Norton	2006	Eggplant, papaya, potato, tomato	Trial data, farmer survey, key informant, published data	Ex ante	Multicountry	Indonesia, Philippines	Economic surplus model
Cabanilla, L. S., T. Abdoulaye, and J. H. Sanders	2005	Cotton	Published data, statistical survey	Ex ante	Country	Mali; also Benin, Burkina Faso, Côte d'Ivoire, Senegal	Linear programming, sensitivity
De Groote, H., W. Overholt, J. O. Ouma, and S. Mugo	2003	Maize	On farm trial data	Ex ante	Country	Kenya	Direct crop loss estimation, economic surplus, closed economy
Falck-Zepeda, J., D. Horna, and M. Smale	2007	Cotton	Published data	Ex ante	Multicountry	Benin, Burkina Faso, Mali, Senegal, Togo	Economic surplus, stochastic simulations
Falck-Zepeda, J., N. Barreto-Triana, I. Baquero-Haeberlin, E. Espitia-Malagón, H. Fierro-Guzmán, and N. López	2006	Potato	On farm data, focus groups	Ex ante	Subcountry	Colombia	Partial budgeting, economic surplus, small open economy, production function input abatement expectations, stochastic simulations
Hareau, G. G., B. F. Mills, and G. W. Norton	2006	Rice	Farmer association data, government data	Ex ante	Country	Uruguay	Economic surplus, monopoly, small country, stochastic simulation
Hareau, G. G., B .F. Mills, G. W. Norton, and D. Bosh	2002	Rice	—	Ex ante	Country	Uruguay	Economic surplus, small country, open economy, stochastic simulation, endogenous adoption

(continued)

Table 5.1 Continued

Authors	Year	Crops	Data type	Time	Scale	Country	Methods
Kambhampati, U., S. Morse, R. Bennett, and Y. Ismael	2005	Cotton	Key informant	n.a.	Subcountry	India	Supply chain analysis
Kolady, D. E., and W. Lesser	2008	Eggplant	Statistical survey	Ex ante	Subcountry	India	Partial budget analysis, willingness to pay, double-bounded dichotomous choice, interval regression, sensitivity analysis, segmented markets, open-pollinated variety hybrid comparison, public-private partnership
Krishna, V. V., and M. Qaim	2007	Eggplant	Statistical survey, field trials	Ex ante	Subcountry	India	Farm survey analysis, trial data analysis, double-bounded dichotomous choice, choice experiment, public-private partnership, open-pollinated variety hybrid comparison
Mamaril, C. B., and G. W. Norton	2006	Rice	Statistical survey (International Rice Research Institute), field trials	Ex ante	Multicountry	Philippines, Vietnam	Economic surplus, large country exporter, small country importer, both open economies, public and private research, national and international
B. Impact on farm and crop sector							
Qaim, M.	1999	Potato	Pilot survey, key informant	Ex ante	Country	Mexico	Economic surplus, small country, closed economy, benefit-cost, sensitivity
Qaim, M.	2001	Sweet potato	Pilot survey, key informant	Ex ante	Country	Kenya	Farm survey analysis, economic surplus, closed economy, cost-benefit analysis, sensitivity

Trigo, E., D. Chudnovsky, E. J. Cap, and A. Lopéz	2002a	Cotton, maize, soybeans	Government data	Ex ante	Country	Argentina	Adoption model, simulation model
Vitale, J., T. Boyer, R. Uaiene, and J. H. Sanders	2007	Cotton, maize	Trial data, published data	Ex ante	Multicountry, focus on Mali	Mali	Economic surplus model, endogenous adoption, risk with lexicographic preferences to capture subsistence-oriented maize production
Pray, C. E., D. Ma, J. Huang, and F. Qiao	2001	Cotton	Statistical survey, key informant	Ex post	Subcountry	China	Farm survey analysis, economic surplus
Qaim, M., and G. Traxler	2005	Soybeans	Pilot survey, key informant	Ex post	Subcountry	Argentina	Farm survey analysis, economic surplus, large open economy, three regions, institutional analysis
Traxler, G., and S. Godoy-Avila	2004	Cotton	Statistical survey, key informant	Ex post	Subcountry	Mexico	Farm survey analysis, economic surplus, small open economy, brief
Traxler, G., S. Godoy-Avila, J. Falck-Zepeda, and J. J. Espinoza-Arellano	2003	Cotton	Statistical survey, key informant	Ex post	Subcountry	Mexico	Farm survey analysis, economic surplus, brief
Yorobe, J. M, and C. B.Quicoy	2006	Maize	Statistical survey	Ex post	Subcountry	Philippines	Farm survey analysis, Cobb-Douglas production function, Heckman model, selectivity bias, producer surplus
C. Impact on crop sector and nutrition							
Zimmermann, R., and M. Qaim	2004	Rice	Statistical survey, key informant, government data	Ex ante	Country	Philippines	Disability-adjusted life years, cost-benefit analysis, simulation

Note: n.a., not applicable; — indicates data type not specified.

Ex Ante Studies

The earliest ex ante sector studies for developing economies were implemented by Qaim. To better represent the semisubsistence nature of producer households in these countries, Qaim applied the Hayami and Herdt (1997) adjustment when modeling the potential impact of virus-resistant and IR sweet potatoes in Kenya and virus-resistant potatoes in Mexico. In both cases he predicted that the transgenic crops would particularly benefit poorer farmers. He warned that limited farmer access to planting material could impede adoption of transgenic potatoes in Mexico. Qaim's explicit reference to seed systems is one of the few in the literature reviewed, though seed and information systems are known to pose major challenges for the adoption of any improved variety, and particularly transgenic crops (Tripp 2001). Massieu et al. (2000) criticized Qaim's assumption that conventional varieties would be completely replaced by the transgenic variety, as well as his assumption that a public delivery system for potato seed would develop in Mexico. In neither of these first two cases did the crop-trait combination considered in the study actually materialize.

The analysis by De Groote et al. (2003) of IR maize in Kenya is a straightforward application of the economic surplus approach supported by comprehensive farm data about maize production practices and on-farm trial data in which crop losses from the pest were measured. The authors highlighted a policy dilemma for the government of Kenya. About 80 percent of the estimated value of crop losses from stem borers in Kenya accrues in the moist transitional and highlands zones, where adoption rates for maize hybrids are greatest and the nation's surpluses are produced. Only 12.5 percent of the value of national crop losses from stem borers occurs in the dry and lowland tropics zones, where the productivity potential for maize is lower. Despite these contrasts, the estimated per-hectare returns from effective insertion of the Bt gene is equal between the zones with lower and higher productivity. This finding was explained by the fact that while maize yields are much higher in the moist and transitional zones, the pest species *Chilo partellus,* against which current Bt proteins were found to be very efficient, inflicts less damage than *Busseola fusca,* for which an effective Bt protein had not been identified. An additional consideration was that the equity impact of developing transgenic materials suited to the low-potential zones could be substantial, since farmers in these zones have fewer alternative sources of income and are generally unable to meet their maize subsistence requirements through on-farm production.

Another early study was conducted by Cabanilla, Abdoulaye, and Sanders (2005), who employed a linear programming model to assess the potential cost to West Africa (in particular Mali) of not adopting Bt cotton. Model parameters were drawn from in-depth farm studies already conducted in Mali, supplemented by secondary data culled from documented experiences in China, Mexico, and

South Africa. On their representative cotton farm, they specify that farmers grow groundnuts and cereals to meet the subsistence needs of their families. Application of the model generates estimates of optimal land allocations, output, farm profit, and whole-farm income. Based on their findings, they then calculate national aggregates and conduct a sensitivity analysis to explore the impacts of technology fees on results. Their results indicate that even with a technology fee, large benefits would be foregone without the adoption of Bt cotton, including more stable farm income. At the level of the fee charged in South Africa, however, groundnuts and non-Bt cotton are no longer produced. They point to key institutional factors that will need to be addressed by policymakers, such as whether the technology will be imported or adapted.

More recent studies treat new crop-trait combinations and introduce other analytical components. Adiyoga et al. (2006) investigated the potential impact of transgenic eggplant, papaya, potato, and tomato in Indonesia and the Philippines in a technical report with multiple authors. Chapters in the report present a combination of estimated partial budgets and economic surplus analyses.

Mamaril and Norton (2006) projected the magnitude and social distribution of benefits from use of Bt *indica* rice in the Philippines and Vietnam, considering cross-country transfers of technology and price changes. One contribution of this study is the comparison between the economic gains from adoption in a small, importing country (the Philippines) and those that accrue to a large, exporting country (Vietnam). Total economic gains were US$329 million in Vietnam and US$270 million in the Philippines in the baseline scenario, of which about two-thirds is earned by producers. Mamaril and Norton find that overall effects of Bt rice adoption on nonadopting countries are not likely to be large. This finding reflects the fact that the amount of rice traded on the world market is small relative to the quantities produced. The authors did not take research costs into account, given the various perspectives from which these costs could be viewed when several countries, and both public and private investors, contribute to product development. Data were provided by the International Rice Research Institute and were assembled from over 700 fields in multilocational trials.

Building on an earlier analysis (Hareau et al. 2002), Hareau, Mills, and Norton (2006) predicted that farmers in Uruguay could expect to earn US$1.82 million in total net present value from adopting HT rice in Uruguay, as compared to earnings of only US$0.55 million for the multinational innovator. According to the authors, this result is disturbing because creating conditions that attract investment is a prerequisite for the use of new technologies in developing countries with smaller markets. On the other hand, stronger IPR regimes would benefit the innovator to the detriment of producers. They propose regional licensing and collaborative research networks to balance these trade-offs.

With respect to methods, Hareau et al. substantially improve on the earlier publications mentioned by incorporating stochastic simulation. Stochastic simulation more adequately depicts the temporal and spatial variability of agricultural production and the uncertainty that farmers experience when adopting new technologies. This technique also permits researchers to pinpoint the factors that contribute the most—as measured in magnitude of effect—to variability in benefits, a finding that can help policymakers choose investments.

Similarly, Falck-Zepeda et al. (2006) and Falck-Zepeda, Horna, and Smale (2007) apply stochastic simulation to estimate the potential sector benefits of IR potatoes in Colombia and Bt cotton in West Africa, respectively. The study conducted in Colombia is unique in at least two respects. First, the authors draw on both data collected from a survey of 78 respondents and data collected from participants in a Farmer Field School. In both cases the data included field verification of integrated pest management (IPM) practices. Second, they estimate expected damage abatement based on historical data and data collected in the field for the study. The authors introduce these data into a stochastic economic surplus simulation which considers the effects of IPM practices and Bt varieties, by variety and for all varieties combined. The authors found that while producers had endured significant field and storage losses within the previous 10 years, these were modest during the survey year, owing to sustained precipitation. Thus, while differences in payoffs between the status quo and the use of IPM practices were zero or negative in a single year, the expectations model demonstrated that producers are likely to gain from investing in IPM over successive seasons. The economic surplus model showed that even considering the variability of field and storage losses and other critical parameters, the use of Bt potatoes in Colombia would generate a positive return to investment in Bt potato research.

The economic surplus model applied by Falck-Zepeda, Horna, and Smale (2007) for Bt cotton in West Africa employs stochastic simulation and relatively conservative assumptions concerning adoption rates, time to release of the new technology, and farm benefits. Several scenarios depict the release of Bt cotton in Benin, Burkina Faso, Mali, Senegal, and Togo. The conservative assumptions employed by the authors explain the relatively small total net benefits they estimate. Nevertheless each of the study countries loses, as a result of declining prices, if it does not adopt Bt cotton while other countries in the world do adopt. The pattern of adoption and the length of the adoption period affect the share of benefits earned by producers as compared to innovators. Thus investments in the institutional "infrastructure" that can support sustained adoption, such as farmer knowledge about the new technology and related practices, and strong marketing channels for cotton inputs and outputs, will be fundamental to earning economic benefits.

Vitale et al. (2007) consider the introduction of both Bt cotton and Bt maize in the cotton production systems of West Africa. This crop combination is a defining characteristic of production in Mali, on which the empirical application of the model is focused. The authors synthesize recent entomological data, drawing on trial data from Burkina Faso. They expand the conventional model of economic surplus by introducing a farm decisionmaking model. Farmer decisions are represented as the outcome of lexicographic preferences, and adoption decisions are endogenously determined within the model. Mali is a small producer of cotton with no market power, so that cotton demand is perfectly elastic; by contrast, maize markets in the country are subsistence-oriented and demand is inelastic. Thus they predict that 81 percent of the social welfare from introducing Bt maize would be earned by consumers, and adoption would be extremely sensitive to the magnitude of the price premium charged as a technology fee. As expected given that most of the crop is exported, adoption of Bt cotton would primarily benefit farmers. Greater profitability of Bt cotton means that adoption could be sustained at higher technology fees. The authors predict that aggregate benefits from Bt cotton would surpass those associated with Bt maize by US$10.3 million per year.

Ex Post Studies

The first ex post study that meets our criteria appears to be the analysis conducted by Pray et al. (2001) of Bt cotton in China. Using a single year of data and the approach proposed by Falck-Zepeda, Traxler, and Nelson (2000a, 2000b), Pray et al. found substantial economic benefits for smallholder farmers. No benefits were earned by consumers because the government bought almost all of the cotton at a fixed price. A consequence of weak IPRs was that farmers obtained the major share of the benefits, with very little accruing to Monsanto or the public research institutions that developed local Bt varieties.

The remaining ex post studies have been conducted for cotton in Mexico and soybeans in Argentina by Traxler and colleagues. Based on survey data for 1997–98, Traxler et al. (2003) and Traxler and Godoy-Avila (2004) find that Bt cotton reduced costs and raised revenues for farmers in the Comarca Lagunera area of north central Mexico, such that "cotton has become a low pesticide crop, benefiting both farmers and residents of the region" (Traxler and Godoy-Avila 2004, 61). Over the two years of the study, they estimated that seed suppliers and innovators earned an average of only 15 percent of the benefits from adoption, while farmers earned the remainder. The authors assert that the risk of crop failure has declined with the use of Bt cotton technology.

In Argentina Qaim and Traxler (2005) combined farm survey data from three regions, institutional information, and secondary data for 1996–2001 to examine the impacts of HT soybeans. Their analysis showed that the United States and Argentina gained while countries that did not produce HT soybeans lost. Farmers in Argentina earned a greater share of the total benefits than U.S. farmers because of weaker IPR protection. One noteworthy detail in their analysis is that some of the model parameters are those estimated under conditions in the United States, which reinforces the perception that soybean producers in Argentina are relatively large-scale, fully commercialized growers. They attribute the success of the technology in Argentina to (1) a suitable agroecology, (2) a strong seed sector that sold large amounts of seed even though IPRs were weak and there were black market sales, (3) adaptive research capability, and (4) a functioning regulatory framework. These factors govern how benefits derived from biotech products introduced in one country "spill over" to other adopting countries.

Conclusions

During the first decade of publications, ex ante studies of sector impacts outweigh ex post studies, and most of these are based on the well-known economic surplus model, with a few methodological adjustments to consider the particular features of developing economies. The dominance of ex ante studies during this period likely reflects the initial interest in whether or not developing economies stand to gain from crop biotechnology. Two important methodological improvements have been the explicit recognition of the temporary monopoly that IPRs confer on the innovator and the use of stochastic simulation to reflect the risk and uncertainty involved in adopting biotech crops.

General Caveats

Alston, Norton, and Pardey (1998) and Scatasta, Wesseler, and Demont (2006), among other applied researchers, have clearly stated the advantages and limitations of the economic surplus approach. The major advantages of the approach are that it is parsimonious with respect to data and can be used to portray the distributional effects of a range of institutional and market structures, thus providing a useful starting point for strategic investment in agricultural research.

The principal disadvantages of the approach relate to incompleteness. The surplus calculated is Marshallian, accounting for price effects but not for changes in the income of farmers. Transaction costs are typically ignored, implying that markets clear and function well. As with any partial equilibrium model, prices and quantities of other commodities produced by farmers are held constant. Effects of technology adoption on input markets are generally not considered. In particular the approach does not account explicitly for returns to land and labor, which are prominent aspects of technology impact. Furthermore farmers are considered to be risk-neutral price-takers who either maximize profits or minimize costs. Positive and negative externalities—such as impacts on the environment, health, and other nonmarket benefits—have not yet been incorporated into these models, although in theory they could be.

Specific Limitations

During the first decade since the introduction of transgenic crops in developing economies, as is the case with other major research questions, sector impact analyses have represented few crops and traits. The assumptions behind the economic surplus approach most closely depict an industry with commercially oriented farmers who buy and sell in well-organized markets and grow their crops under relatively homogeneous conditions. This depiction is unrealistic for most farmers in developing economies, and particularly those who produce staple foodcrops.

The quality of the underlying data is crucial for the validity of the results. In general, reliable cross-sectional, time-series data to support sector analyses of transgenic crops are not yet available in most developing economies. At present such data are probably too costly to assemble, maintain, and disseminate publicly given the information infrastructure found in most of these countries. The databases employed by researchers in China and India are exceptional. In the United States, as a point of comparison, extensive, publicly funded surveys have been conducted continually, such as the annual Agricultural Resource Management Survey, on which many of the detailed analyses are based. "Pure" ex ante analyses (with no field observations) are even more limited, since all model parameters must be projected based on expert interviews and existing secondary data.

One way in which researchers have compensated for the lack of large cross-sectional, time-series data has been to expand existing data from both primary and secondary sources using stochastic simulation. These tools assume special significance when technologies are grown by farmers in heterogeneous production environments for uncertain markets, where location and year-specific effects on productivity can generate large coefficients of variation in model parameters, including farm profits, adoption rates, and prices. If the number of input suppliers is small or markets must be segregated, risk and uncertainty in the market channel may be somewhat higher in the case of transgenic crops relative to other new crop varieties. Markets have not been segregated in the developing economy cases that have been documented so far.

Given the salience of issues related to biosafety regulations, supply channel performance, and industrial organization in the development and diffusion of transgenic crops, quantitative analyses of these issues are particularly needed. In addition issues such as effects on health, the environment, and poverty have not been adequately addressed. One explanation for the rarity of such studies is that these topics may be less amenable to analysis with conventional, applied research methods.

Chapter 6

Impacts on International Trade

Approaches

Conducted at a higher level of aggregation than sector studies, trade studies assess the effects of the adoption of transgenic crops in multiple countries and often across multiple sectors. Instead of focusing on the economic benefits of a transgenic crop within the national agrofood industry and their distribution among different agents, trade studies analyze the international consequences of transgenic crops for adopting and nonadopting countries, including exporters and importers. Assumptions invoked in trade models relate to the nature and magnitude of productivity changes, the organization of markets, and regulatory frameworks.

Trade studies are particularly relevant in this topic area because the four major transgenic crops on the market (soybeans, maize, cotton, and canola) are also major, internationally traded commodities. Trade studies of transgenic crops differ from those of other agricultural products mainly because of the presence of trade-related regulations specific to biotech products. Moreover the issue of market segregation between products made with transgenic and nontransgenic ingredients is increasingly important because consumer acceptance of biotech foods differs across countries. The international scope of these studies often leads to a discussion of the political economy of agricultural biotechnology, as evidenced by the common practice of separating countries into three groups. The first group is composed of major exporting countries that have embraced transgenic crops. The second group includes major developed nations that import transgenic crops, regulate the trade and marketing of transgenic products, and tend to be less accepting of transgenic crops. The third group is composed of developing economies that are potential importers of transgenic crops but do not currently produce them.

Nielsen, Thierfelder, and Robinson (2003) initially reviewed empirical trade studies of the introduction of transgenic crops, but a number of new studies have been published since then. Aside from purely theoretical treatments (Plastina and Giannakas 2007) or more cursory forms of forecasting (Brookes and Barfoot 2006), applied trade studies can be divided into three sets according to the methods used by the authors. Although all three sets are based on the use of ex post international trade data, virtually all studies consist of ex ante analyses of the potential international trade implications of adopting transgenic crops.

The first set includes studies that use existing data on international trade to support a descriptive or accounting analysis of trade-related issues driven by the adoption of transgenic crops. Recent patterns of exports of transgenic crops (soybeans, maize, cotton, and canola) by one or more countries are analyzed with bilateral trade data. Countries under study either now produce or are considering the production of transgenic crops. Authors seek to assess the economic effects of the association of transgenic crop adoption with changes in market access. Even when these studies are based on a conceptual framework, their empirical applications are not grounded in an explicit model of international markets.

The other two sets of studies represent two major categories of applied trade models. In the first category partial equilibrium models are applied to analyze one or several sectors of the economy in a few countries, focusing on particular vertical or horizontal linkages. These studies have the advantage of flexibility, enabling the representation of a complex array of institutional and market policies. Their principal disadvantage is that they do not take into account linkages with multiple sectors. Nor do they consider specific regulations affecting bilateral trade relationships with "sensitive" importing countries (countries in which transgenic products are not well accepted by consumers).

The second category employs multicountry, general equilibrium models. These models provide a consistent and comprehensive structural representation of the economy and of international trade linkages. However, they are less conducive to representing specific policies and institutional arrangements because they are highly aggregated and based on important assumptions about the market. Descriptors for all three sets of studies are shown in Table 6.1.

Descriptive Studies

Two studies focus on the commercial risks that could result from importers' regulation of transgenic products. Paarlberg (2006) assessed the potential export losses that might be incurred if African countries adopted current transgenic crops, using 2003 data on bilateral trade from 16 countries exporting to Europe. He showed that the commercial risks associated with the use of transgenic crops are minimal both

Table 6.1 Study descriptors: Impact of transgenic crops on international trade

Authors	Year	Crops	Data type	Scale	Country	Methods
Anderson, K., and L. A. Jackson	2005	Coarse grains, oilseeds, rice, wheat	Trade	Global, focus on Sub-Saharan Africa	n.a.	Applied general equilibrium model
Anderson, K., L. A. Jackson, and C. P. Nielsen	2004	Rice	Trade	Global, focus on Asia	n.a.	Applied general equilibrium model
Anderson, K., and S. Yao	2003	Cotton, maize, rice, soybeans	Trade	Global, focus on China	China	Applied general equilibrium model
Anderson, K., C. P. Nielsen, S. Robinson, and K. Thierfelder	2001	Gm	Trade		n.a.	Applied general equilibrium model
Anderson, K., and E. Valenzuela	2007	Cotton	Trade	Global, emphasis on Sub-Saharan Africa	n.a.	Applied general equilibrium model, Global Trade Analysis Project, removal of subsidies and tariffs and genetically modified crops
Anderson, K., E. Valenzuela, and L. A. Jackson	2008	Cotton	Trade	Global, emphasis on Sub-Saharan Africa	n.a.	Applied general equilibrium model, partial budgets
Annou, M. M., F. H. Fuller, and E. J. Wailes	2005	Rice	Trade	Global	n.a.	Arkansas Global Rice Model
Díaz Osorio, J., R. Herrera, J. Valderrama, and J. L. Llanos Ascencio	2004	Maize	Farmer association, company data	Country	Chile	Partial budget analysis, sensitivity, partial equilibrium analysis
Berwald, D., C. A. Carter, and G. P. Gruère	2006	Wheat	Trade	Global with emphasis on Canada and United States	n.a.	Partial equilibrium analysis
Elbehri, A., and S. Macdonald	2004	Cotton	Trade	Global, focus on West and Central Africa	Benin, Burkina Faso, Cameroon, Central African Republic, Chad, Côte d'Ivoire, Mali, Togo	Applied general equilibrium model

(*continued*)

Table 6.1 Continued

Authors	Year	Crops	Data type	Scale	Country	Methods
Felloni, F., J. Gilbert, T. I. Wahl, and P. Wandschneider	2003	Grain	Trade	Global, focus on China	China	Applied general equilibrium model
Frisvold, G. B., J. M. Reeves, and R. Tronstad	2006	Cotton	Trade	Global, focus on China	China	Three regions, output price endogenous model calibrated to 2001, monopoly
Gruère, G. P., A. Bouët, and S. Mevel	2007	Genetically modified crops	Trade	Multicountry	n.a.	Applied general equilibrium model
Hareau, G. G., G. W. Norton, B. F. Mills, and E. Peterson	2005	Rice	Trade	Global, focus on Asia	Bangladesh, China, India, Indonesia, Japan, Philippines, Thailand, Vietnam	Applied general equilibrium model
Huang, J., R. Hu, H. van Meijl, and F. van Tongeren	2004	Cotton	Statistical survey, field trials, trade	Global, focus on China	China	Applied general equilibrium model, modified Global Trade Analysis Project
Langyintou, A. S., and J. Lowenberg-DeBoer	2006	Cowpea	Trade	Multicountry	n.a.	Spatial and temporal price equilibrium model, mixed complementary programming framework, differential interest rates
Moschini, G.	2001	Multiple	Trade	Global	n.a.	Review, findings, world trade model, brief
Moschini, G., H. Lapan, and A. Sobolevsky	2000	Soybeans	Trade	Global	n.a.	Three-region model, monopoly, simulation, partial equilibrium model

Nielsen, C. P., and K. Anderson	2001a	Maize, soybeans	Trade	Global, selected countries	n.a.	Applied general equilibrium model
Nielsen, C. P., K. Thierfelder, and S. Robinson	2003	Maize, soybeans	Trade	Global	n.a.	Multiregion computable general equilibrium model
Nielsen, C. P., and K. Anderson	2001b	Maize, soybeans	Trade	Global	n.a.	Applied general equilibrium model
Nielsen, C. P., S. Robinson, and K.Thierfelder	2001	Maize, soybeans	Trade	Global	n.a.	Applied general equilibrium model
Paarlberg, R.	2006	Genetically modified crops	Trade	Multi-country	n.a.	Review, calculations with secondary data
Smyth, S., W. A. Kerr, and K. A. Davey	2006	Canola, maize, soybeans	Trade	n.a.	China, Colombia, Egypt, Indonesia, Korea, Mexico, Pakistan, Taiwan	Graphical model and secondary data
Sobolevsky, A., G. Moschini, and H. Lapan	2005	Soybeans	Trade	Global, focus on Argentina, Brazil, United States	Argentina, Brazil	Partial equilibrium world trade model, segregation, import bans
Wu, F.	2006	Maize	Trade	Global	Argentina, China	Integrated assessment model, sensitivity analysis, probability density functions, export losses, benefits in market effects, animal health, human health

Note: n.a., not applicable.

in absolute value and relative to total export value. He concluded that the observed fear of export losses in these countries is overrated. Smyth, Kerr, and Davey (2006) developed a partial equilibrium framework to investigate the effects of import restrictions in selected markets and applied it to the case of major transgenic crops in Canada and the United States. The authors used bilateral export data for the periods before and after the introduction of these crops. They showed that these two countries did not suffer from any observable reduction of export values in the crops. Introduction of transgenic crops resulted only in the diversion of trade from traditional importers who were reluctant to buy transgenic crops toward more open markets. Their finding refutes the claim by some activist groups that transgenic exporters experienced huge losses due to market restrictions.

On the benefit side, for three big exporters of maize (the United States, Argentina, and China), Wu (2006) derives the commercial benefits of using Bt corn to increase market access under different harmonized international standards for mycotoxins. The Bt toxin expressed in Bt maize naturally reduces the presence of three mycotoxins (fumonisin, aflatoxin, and deoxynivalenol) and therefore could help exporters access markets with stringent safety standards. Wu assumes potential rejection rates at the frontier and multiplies potential rejected volumes by the efficacy rates of Bt maize in reducing these toxins. Results show that Bt maize could help eliminate 22 percent and 9 percent of losses due to tests on aflatoxins and fumonisins, respectively, under highly restrictive international standards such as those of the EU.

Partial Equilibrium Models

The partial equilibrium approach is applied in several articles by Moschini and Sobolevsky. Moschini, Lapan, and Sobolevsky (2000) examined the welfare effects of HR soybeans and various IPR scenarios in a partial equilibrium, three-region model composed of Argentina, Brazil, the United States, and the ROW (rest of the world). They assumed that the technology reduces costs and in one case increases yields. Their results suggest that the United States gains most, with the innovator capturing the largest share of the gains. U.S. farmers gain too, but not if the innovation enhances yields. Technology spillover to Latin America erodes the competitive position of U.S. soybean producers. With weak IPRs in Latin America, profits from sales of the new technology just offset the loss of U.S. producer welfare. Consumers in every region gain from adoption of HR soybeans.

Based on secondary data and the findings from the application of this model, Moschini (2001) underscored the role of disparities in IPRs across countries in the distribution of benefits from adopting biotechnology innovations. IPRs are perceived to be necessary to address market failures in research and development of improved germplasm which lead to some efficiency losses. Yet the exercise of market power lowers benefits from adoption because the innovation is not used "as is socially desirable. . . . Consumers gain less, and farmers' welfare is reduced. Innovators gain more" (Moschini 2001, 113). He added that consumer resistance, labeling, and market segregation complicate the economic evaluation of these technologies.

Sobolevsky, Moschini, and Lapan (2005) used a partial equilibrium trade model but included product differentiation and the costs of identity preservation in segregating markets. This approach generates some unexpected findings and new hypotheses. The authors examine the trade and welfare effects of HT soybeans on Argentina, Brazil, the United States, and the ROW. Consumers in the importing region view transgenic soybeans and products as weakly inferior substitutes. Sobolevsky, Moschini, and Lapan (2005) find that in a world where no segregation is feasible, the long-run equilibrium is worldwide adoption. This leads to lower prices, with the United States leading in exports and all regions and economic agents gaining except U.S. farmers. When segregation technology is available at a cost, the United States emerges as the only region with partial adoption, and all other regions specialize in HT soybeans. Output subsidies cause welfare reduction to the United States and only the ROW gains because it offsets the distorted prices caused by monopoly in the innovation. With import bans by the ROW and Brazil, Brazilian farmers would benefit and the ROW could benefit if segregation costs were not too low.

Annou, Fuller, and Wailes (2005) used a preexisting international model of rice to model the effect of the adoption of drought-tolerant, transgenic rice in multiple countries. First they considered the commercial diffusion of transgenic rice in countries categorized as early adopters according to their need and resources in biotechnology, comparing the effects in terms of two yield levels and three adoption rates. They then compared these results to similar scenarios for a broader group of countries. Their results show that transgenic rice would increase production, decrease price, and expand rice consumption, for the benefit of consumers. Producer welfare is small or negative because of an immiserizing growth effect. Early adopters are bound to gain, to the detriment of producers in less-developed countries because of the price decline. Their analysis clearly benefited from the use of a comprehensive and detailed international commodity model, but the implications of their results are limited by the fact that they include no trade regulation on transgenic rice.

Berwald, Carter, and Gruère (2006) used another partial equilibrium trade model to study the global and regional welfare effects of adoption or nonadoption of HT wheat in Canada and the United States. In addition to these two countries, Argentina and a region grouping major wheat producers in the developing world are included in the simulation. The model features heterogeneous consumers (differentiated by region, type, and taste), segregation costs, and the effects of two types of labeling regulations for biotech food in major trading countries. Labeling affects world prices and shares of consumers purchasing transgenic or nontransgenic wheat. Berwald, Carter, and Gruère (2006) found that Canada and the United States will face significant welfare losses if they do not adopt transgenic wheat at the same time as Argentina and other wheat producers do. Their results also showed that most gains from the adoption of transgenic wheat would occur in the developing world, but all adopting countries gain despite the barriers to entry in sensitive importing countries. They conclude that the decision of North America to reject transgenic wheat "supports the misleading argument that market segregation is absolutely impossible and that sensitive markets should dictate choices over agricultural biotechnology" (Berwald, Carter, and Gruère 2006, 445).

Langyintuo and Lowenburg-DeBoer (2006) used a regional spatial and temporal trade model to assess the effects of Bt cowpea adoption in West and Central Africa. The model includes transport costs, and this allows for measuring trade diversion associated with the adoption of Bt cowpea in one or more countries in the region. It also accounts for the different periods in the growing and marketing season for cowpeas. The results show that Bt cowpea adopters gain, while other countries may not gain as much because nonadopting producers will lose. Three different scenarios of adoption and yield levels were simulated, and a sensitivity analysis was conducted on the adoption rate in Nigeria. Findings concur with the general evidence that increased adoption is beneficial for consumers everywhere, but less so for producers.

Last in this group, Frisvold, Reeves, and Tronstand (2006) built a three-region, partial equilibrium model of the cotton market to measure the international market effects of Bt cotton introduction in China and the United States. They separated the effects of adoption in each of these two countries and also calculated the scenario with adoption in both countries. They found that the gains associated with adoption in both countries almost perfectly equal the sum of the gains obtained in the other scenario. They also found that the net global producers' surpluses are US$444 million, of which China gains US$423 million and the United States US$179 million. They show that the world price decreases by US$0.014 per pound with Bt cotton adoption, which results in producer surplus losses of US$349 million in nonadopting regions.

General Equilibrium Models

Fifteen distinct published articles apply general equilibrium models. All 15 articles use a modified version of a computable general equilibrium model based on the GTAP database (Hertel 1997) that includes vertical and horizontal linkages in the economy. This modeling framework is used to examine the effects of transgenic technology adoption on multiple sectors and regions. The articles differ in their assumptions about the productivity effects of the technology and adoption rates, and according to the scenarios they depict concerning trade policies, consumer perceptions, and the structure of the nontransgenic/transgenic market chain. These papers can be divided into four groups, defined by their successive contributions to improvement in modeling.

Two papers led the way in evaluating the economywide international effects of the introduction of transgenic crops. Nielsen, Robinson, and Thierfelder (2001) studied the introduction of transgenic soybeans and maize in seven regions. They modeled the technology with a 10 percent Hicks-neutral productivity shift of primary factors, with costless segregation of transgenic and nontransgenic food and consumer price sensitivity differences by adjusting demand elasticities of substitution between transgenic and nontransgenic. They showed the effect of changing consumer acceptance on the different market factors in developing countries. In parallel, Nielsen and Anderson (2001a, 2001b) provided a global study of the introduction of transgenic soybeans and maize in a larger number of countries and regions, using a 5 percent Hicks-neutral productivity shift on factors and intermediate consumption to model the effect of the technology.[1] They simulated scenarios that show the effects of a 25 percent decrease in consumer demand in sensitive countries or an EU ban on imports of transgenic food.

A second group of papers provided slight refinements to the methodology. Stone, Matysek, and Dooling (2002) focused on Australia within a multiregion world and modeled the introduction of transgenic maize and soybeans based on updated data, using more accurate productivity shifts (6 percent for oilseeds and 7.5 percent for others), more realistic national adoption rates, and consumer demand changes, as well as regulatory costs. Nielsen, Thierfelder, and Robinson (2003) used their former model to study the effects of consumer acceptance on the benefits of biotech food products by combining the two modeling options pursued before: they varied price sensitivity and added utility shifts (consumer acceptance) to show how consumer acceptance can affect results and improve the segregation of biotech and non-biotech food products. Anderson and Yao (2003) focused on China and applied the same method to cotton, maize, and soybeans, with an additional scenario that eliminates the Chinese voluntary export restraint on textiles. Anderson and Jackson (2003) used the productivity shifts of Stone, Matysek, and Dooling

[1]Nielsen and Anderson (2001a, 2001b), and the authors of papers following them, believe that 5 percent is a conservative estimate of the potential of the technology. Yet this opinion is not shared by all. For example, Felloni et al. (2003) show that in order to assure grain self-sufficiency, plant biotechnology would have to result in an annual 4 percent productivity shift, which they believe is very unrealistic.

(2002) and considered various trade restrictions in the case of the introduction of transgenic soybeans and maize to focus on the political and economic implications of EU-U.S. regulatory differences.

Third, van Meijl and van Tongeren (2004) provided a study of the introduction of biotech food products in the EU and United States, with a change of methodology. The change was made in response to a significant criticism of the Nielsen and Anderson (2001a, 2001b) approach. They replaced Hicks-neutral shifts by factor-biased productivity shifts for cereals and introduced technology spillover. The authors also included a more realistic representation of the EU's Common Agricultural Policy (CAP) by including the isolation of EU countries from world prices. Despite the ongoing reform of CAP, several of its programs contribute to the disconnection between world prices and EU prices. Van Meijl and van Tongeren show that because they did not take this situation into account, Nielsen and Anderson (2001a, 2001b) overestimated the negative welfare effects of an EU import ban on EU consumers.

The fourth and largest group of published studies focused on specific regions and/or commodities. Authors employ more realistic assumptions with mixed Hicks-neutral and factor-biased productivity shifts and additional layers of complexity to represent international markets. Most of these papers deal with Bt cotton (using ex post field data) and/or transgenic rice, focusing on balancing their productivity gains with the effects of biotech food regulations on importers.

Huang et al. (2004) analyzed the effects of transgenic cotton and rice introduction in China. They use farm-level data from three regions of China to formulate assumptions about the expected national productivity gains with transgenic cotton, accounting for changes in yield and the costs of pesticides, seeds, and labor with the technology. Their factor-biased productivity shift is a real improvement over other studies. They add labeling costs, loss of demand in export markets, and model adoption in a dynamic framework, but without adoption of these crops in any other country. Their results show that China can continue to benefit from an extended adoption of Bt cotton, but that it will benefit even more from the introduction of GM rice, whose formal approval has been postponed by regulatory authorities in the past few years.

Elbehri and MacDonald (2004) evaluated the potential effects of transgenic cotton in West and Central Africa based on a careful analysis of productivity effects in the region (using farm and national budgets) and comparing various productivity shifts. The amplitude of the Hicks-neutral productivity shock is determined by changes in factor use. They employ a series of national production budgets for cotton to compute the weighted average of the effects of gains in yield and reductions in the costs of seeds, pesticides, and labor on the total factor productivity in each country. They also model the productivity decline without new technology in West and Central Africa to account for the current trend observed in the region.

They find that West and Central Africa not only would gain from the adoption of Bt cotton but would suffer significant losses without adoption.

Similarly Anderson and Valenzuela (2007) evaluated the effect of Bt cotton introduction in Sub-Saharan Africa (SSA) with or without World Trade Organization trade reform. First they assess the effects of removing subsidies and tariffs on cotton. They then simulate the introduction of GM cotton. The results show that SSA would gain appreciably from both changes. Net farmer income increases by 10 percent if farmers adopt Bt cotton and it is reduced by 7 percent if they do not adopt. Their findings also demonstrate that the use of Bt cotton would be more beneficial if it is adopted simultaneously than if it follows trade reforms.

On the food side, Anderson, Jackson, and Nielson (2004) evaluated the effects of transgenic rice introduction (Bt and Golden Rice) in developing countries, with updated assumptions about factor-biased productivity and potential moratoria in Europe and Southeast Asia. They show that Golden Rice could provide a much bigger boost to countries adopting it owing to its assumed effect on overall labor productivity in all sectors. Anderson and Jackson (2005) used the same framework to focus on the introduction of GM maize, rice, soybeans, and wheat in Australia and New Zealand under various trade scenarios.

Hareau et al. (2005) evaluated the effects of three different transgenic rice products (Bt, HT, and drought-tolerant) in eight countries of Asia and five other regions. They use factor-biased productivity shifts, accounting for intranational differences in land type and distinguishing the effects of these different events given favorable land and three types of unfavorable land, thereby providing a convincing approach to productivity modeling. Their results show that if the benefits of the three technologies are similar overall, the distribution of benefits will depend highly on the particular trait. The main limitation of this study is that the authors do not consider the effect of importers' policies.

Last in this group, Gruère, Bouët, and Mevel (2007) built on a number of the foregoing approaches to achieve improved representation of productivity effects and market restrictions. They focused on cotton, maize, rice, soybeans, and wheat in Bangladesh, India, Indonesia, and the Philippines, but they also included adoption of these crops in other countries. The factor-biased productivity shocks they assume for the four countries are based on interviews with local experts combined with a spatial analysis of abiotic constraints (drought, salinity). In addition their model separates transgenic and nontransgenic products within each sector, helping them to model the short- and long-run effects of import approval policies. The approach also explicitly incorporates the effect of labeling policies, by allowing only products intended for intermediate consumption to enter sensitive countries. Their results show that transgenic rice would be the most beneficial crop for these four countries, and that the gains from adopting transgenic crops largely exceed any type of trade losses due to market restrictions. They also derive the opportunity costs of segregation for many countries and show that segregation will be more valuable for importers with strict regulations than for exporters that adopt transgenic crops.

Conclusions

During the past seven years, trade studies have progressively improved in terms of their representation of international economic realities. As information and regulations have improved, these studies have become increasingly focused and have relied on more detailed representations of transgenic crop technology and the complexities of the international commodity market. Although the main policy question has remained the same, the subquestions examined by authors have evolved from the general effects on price and welfare to the opportunity cost of segregation, the importance of trade diversion, or the combination of transgenic crop adoption with other changes in economic policy.

This review of methodologies reveals a variety of approaches, from simple calculations based on bilateral trade data to economywide trade models that incorporate complex relationships. Each approach has its own advantages and tends to complement the others. Instead of increasing specialization, constraints may have encouraged researchers to move toward a partial convergence in methodology. For example, most studies that use partial equilibrium models are based on a representation of aggregate trade in which prices clear via the equilibrium of supply and demand. Such studies cannot be employed to assess the effects of a regulation on bilateral trade flow from a transgenic exporter to a restrictive importer, or the effect of trade diversion. Such approaches would benefit from incorporating bilateral trade relationships into the studies, as in the general equilibrium studies. Computable general equilibrium models, on the other hand, are limited by excessive aggregation of sectors and regions. These models have been used with increasingly more detailed and realistic intrasector specifications of the technology and/or regulations, following partial equilibrium models.

Despite the variety of approaches, studies consistently support three main findings. First, they illustrate the importance of "first-mover" advantage. Countries that do not adopt transgenic crops lose if they stay behind. Second, a number of studies highlight the risk of immiserizing growth that occurs when yield-increasing technologies are adopted in markets with inelastic demand. Under these circumstances, growth translates into a price decline that benefits consumers to the detriment of adopting producers. Third, many studies demonstrate that in developing economies the potential export losses resulting from the adoption of transgenic crops are outweighed by the potential gains from productivity enhancement. Of these three findings, only the last is specific to transgenic crops compared to other productivity-enhancing agricultural technologies. Yet this last finding is of particular interest: how can one explain the fear of export losses expressed by governments in countries with commercial risks that are in fact quite limited? Political economic studies, and perhaps the creation of a new field of research, would be necessary to respond to this question.

General Caveats

Each group of studies suffers from particular caveats, but the various approaches share a number of limitations. First, because these are ex ante studies, they are based on assumptions that are naturally uncertain. In comparison with the sector studies discussed in Chapter 5, the treatment of uncertainty regarding multiple assumptions and parameters remains relatively unsophisticated. Because of the complexity of some of these simulation models, sensitivity analysis is accomplished with only a few points on a few parameters. Only one partial equilibrium analysis used a stochastic representation of parameters. Second, because the macroeconomy is their focus, the studies rely on a number of important assumptions concerning aggregation. Some of these may be problematic, such as the aggregation of nonhomogenous countries, sectors, types of crops, or varieties. Third, most models are based on the assumption of perfectly competitive markets, which is likely to be unrealistic given the structure of the commodity supply chains, particularly in the input sector. Fourth, there has been no explicit effort so far to reconcile results from trade studies with the reality of markets in developing countries. Market integration is not perfect in most developing nations, and many of these countries maintain price-distorting policies. Therefore price transmission is imperfect, the distribution of rents is asymmetric, and producers or consumers may not be directly affected by international changes.

Finally, apart from descriptive analysis, we found no ex post empirical study of the effects of transgenic crops on the international market. The major constraint is the lack of data on the transgenic content of bilateral trade flows. There may be some data on identity-preserved, nontransgenic trade flows, but no data are available on trade flows of transgenic products. Nevertheless one feasible approach would be to examine international exchanges of commodities that may contain transgenic products, excluding trade volumes of nontransgenic crops and multiplying adoption rates by export or import quantities of available transgenic products. This approach could help to gauge the effect of introducing transgenic crops on prices or other trade variables, in the presence and absence of trade-related regulations.

Specific Limitations

Descriptive studies can provide only limited insights from quantitative conclusions, and they do not consider price effects. Some of these studies may provide useful first steps, but the quality of these analyses depends critically on the source of the secondary data on which they are based. Bilateral trade flows are well reported by countries in the Organisation for Economic Co-operation and Development (OECD), but this is not the case for a number of developing countries, particularly those in SSA. Databases that are generally available, such as the United Nations' Comtrade or FAOSTAT, tend to report numbers as they accumulate, without the assurance of consistency across data points.

Only a few published articles use partial equilibrium simulation models to evaluate the introduction of transgenic crops, and even fewer focus on developing countries. Data constraints also affect the utility of these studies. The more realistic the data, the more realistic the model parameters used to predict the effect of regulations in international markets and more specifically in developing nations. For instance, it is difficult to obtain data about the demand for transgenic or nontransgenic crops given the product-specific nature of labeling regulations in major OECD importers. Figures for costs of segregation are based mostly on studies conducted in industrialized economies, which are not realistic for nonindustrialized economies. Related to this problem is the fact that analysts need to build customized models of existing regulations and segregated markets for transgenic and nontransgenic crops. Unfortunately simulations conducted in partial equilibrium frameworks rely on relatively simplistic assumptions about the adoption and productivity effects of the technology.

The progressive improvement in applied general equilibrium modeling in the published literature has resulted in deflation of the computed welfare effects associated with the introduction of the technology. Initially the world was modeled as achieving gains amounting to US$10–12 billion with the introduction of transgenic soybeans and maize, but more recent models estimate these gains to be about US$4–7 billion with the same transgenic crops. The accuracy of the results has been improved by more realistic assumptions concerning productivity shifts and adoption rates, as well as the availability of the updated GTAP database, with more realistic economy linkages on the one hand and segregation, demand, and trade-related regulations on the other.

Yet several key methodological issues remain to be resolved in these models. The first concerns the productivity changes brought about by adoption of transgenic crops. Despite improvements in modeling productivity, a fuller articulation is needed that considers regional differences, effects on labor markets, heterogeneous land types, and, in particular, seed prices. Most studies do not make an effort to adjust for the aggregation of sectors in the GTAP database. For example, to model the introduction of maize, technology shifts were introduced into the cereal sector of GTAP. Yet the cereal sector includes not only maize but other major crops, such as barley and sorghum. Similarly the oilseed sector of GTAP was used to model the introduction of transgenic soybeans, ignoring crops such as rapeseed and mustard. The GTAP plant-based fiber sector is used to represent only cotton, despite the fact that jute and linen can account for a significant share of production in this sector. Adoption rates are exogenous and somewhat arbitrary. Modeling adoption as endogenously determined in a dynamic framework would improve the utility of these models.

A second issue involves the representation of markets. There is no effort to model market imperfections in the input sector. Only two studies include the costs of segregation for nontransgenic products. These analyses do not completely model the actual situation, which includes pure nontransgenic and organic trade as compared to mixed-commodity (transgenic and nontransgenic) trade. Consumer acceptance and labeling effects could be better addressed.

Trade constitutes a third issue. Trade-related regulations of biotech food, particularly labeling and import approval regulations, need better treatment. Most articles model regulations as import bans in the EU, Japan, or South Korea when in fact these countries do import large volumes of undifferentiated soybeans and/or maize from countries producing transgenic crops for animal feed and nonfood uses. Only one study separates the short-run effect of import approval regulations from the long-run effects of importers' policies, using a trade filter to model mandatory labeling regulations. In addition markets for unauthorized seeds of transgenic crops have not been represented, although certain countries are known to have only weak controls in place at their borders.

These limitations call for improvements in the methodology. While applied general equilibrium evaluations can be improved through the use of more realistic field and regulatory data, some of the issues of interest to policymakers may be difficult if not impossible to model within already complex macroeconomic models of international trade.

Chapter 7

Conclusions and Future Directions

Findings

Since the first transgenic crops were released to farmers in 1996, documented experience with these crops in developing agriculture has accumulated but remains narrow in scope. Only a few crop-trait combinations have been examined in a limited number of countries, and the period of adoption is brief. Focusing on applied economics methods, this review has examined the evidence that is based on the analysis of empirical data and has been published in peer-reviewed articles. Understanding research protocols and analytical approaches in this first decade of study is important for two reasons. First, it improves our comprehension and interpretation of findings. Second, it facilitates methodological advances by applied researchers in the next generation of studies. Improved methods will enhance the quality of information about the economic impact of biotech crops in developing countries.

Studies have been grouped into categories that correspond to research questions and fields of analysis. The first category of study analyzes the adoption of biotech crops and its impacts on farmers. The second treats the attitudes of consumers toward products made with transgenic ingredients. The third set considers the impact of biotech crops on a given industry or sector, which is composed of both producers and consumers. Studies of the impacts of transgenic products on international trade constitute the fourth category. Each field of analysis is associated with certain types of data, analytical models, and policy implications.

In this chapter, under each field of analysis, we recap our findings and recommend future research directions.

Farmers

Conclusions

Literature about the economic impact of transgenic crops on farmers is the most extensive among the four topic areas examined; it is also especially informative because almost all of it is ex post. In contrast to ex ante analysis of potential impacts, ex post research documents actual patterns of adoption and impacts.

During the first decade of their use by smallholder farmers in developing economies, peer-reviewed research has indicated that, on average, transgenic crops do provide economic advantages for adopting farmers. However, several general caveats are useful to remember when interpreting the findings reported in this initial literature. A number of specific limitations have also been identified in this review.

The first general caveat is that only a limited range of transgenic crops has been studied because few have been released in developing countries. Studies of Bt cotton, which has unique economic and agronomic properties, dominate the literature; a few country case studies also dominate the Bt cotton story. Thus we should be careful not to generalize from these experiences to other crop-trait combinations and contexts. Similarly there are relatively few *different* authors publishing case studies in peer-reviewed international journals, and there is also a wide range of quality among the journals publishing the research.

A second general caveat is that averages mask considerable variation. The magnitude of the economic advantages varies substantially according to the nature of the cropping season and the geographical location of the study. This would be the case whether or not the seed introduced were transgenic, but the variation is particularly pronounced for IR crops. Variability in crop yields and profitability reflects the reliance of agricultural production on uncertain weather conditions and pest pressures, combined with the heterogeneity of farmers, farming systems, and farm-related institutions. Not all farmers will benefit from IR crops in every cropping season, and this variability is difficult to capture in cross-sectional data collected in single locations.

Related to this caveat is a third: the length of the period over which adoption and impact are observed can dramatically influence the conclusions drawn by researchers. Some success stories are episodic; others are not apparent until years have passed. The impacts we are able to observe also depend on the point along the adoption path that is studied. During the initial years of adoption, it makes sense that researchers have focused on the relative profitability of transgenic crops; if transgenic crops are not advantageous for farmers, they will not adopt them and there will be no measurable impact of any kind. Only after farmers have planted transgenic crops for a number of years can we assess empirically the effects of adoption on poverty, inequality, health, and the environment.

Specific methodological limitations in the first generation of studies include selection bias, measurement errors, and biased estimation procedures. Most published studies exhibit potential bias associated with study placement, selection of farmers through a company extension program, and/or self-selection of certain types of farmers into the adopting group. Comparisons of input use and yields between adopters and nonadopters have generally been made with subjective approaches such as farmer recall rather than more objective protocols, contributing to imprecise estimates. Explicit treatment of risk and uncertainty has been rare. The endogeneity of decisions regarding variety choice and input use has not always been taken into account in econometric analysis, which could bias the estimated effects of adoption on productivity and pest damage.

Future Research Directions

The limitations cited earlier should be overcome in the next generation of studies. Authors have sought to address selection bias in various ways, although methods to minimize selection bias are thoroughly explored in the broader economics literature, and techniques such as randomized experiments and treatment models should be brought to bear more rigorously on this topic. The political controversies surrounding transgenic crops render it more difficult to randomize samples or design "experiments." Although desirable from a statistical standpoint, these approaches are more likely to be socially unacceptable in the case of transgenic crops. Obtaining the consent of randomly selected respondents or farming communities may not be feasible. At the same time, political controversies underscore the need to document impacts as neutrally as possible. In addition, although placement bias and bias associated with failure to account for the characteristics of early adopters can be mitigated through careful study design, researchers will continue to face challenges in identifying the appropriate counterfactual for use of specific varieties and practices.

As the range of biotech crops released in developing countries expands, a wider range of farm modeling frameworks will be needed. Stochastic simulation and damage abatement models have constituted important methodological improvements in this literature, but the absence of more complete decisionmaking models—such as risk theoretic frameworks, household models that consider market imperfections, and dynamic models that incorporate learning processes—is noteworthy.

Beyond these limitations, it is evident that the topics of investigation must be expanded considerably in the next generation of studies in order to shed light on the debates that most interest policymakers and the public. First, the features of transgenic seed that differentiate it from other seed should be examined more closely. For example, a unique aspect of transgenic crops is the way that knowledge about them is disseminated and distributed among farmers and other actors in society. The dynamic evolution of farmer knowledge about transgenic crops, and how knowledge is transferred to and among farmers, is an important but neglected component of the economic analyses published so far. Generally, knowledge is expensive in atomistic, dispersed rural societies without advanced information technology.

Some proponents contend that embodying the insecticide in the seed removes much of the uncertainty or risk associated with the timing and intensity of chemical applications. Therefore use of IR transgenic crops is simpler for poorly informed or less literate farmers. This perspective is shortsighted, however. With traits such as insect and disease resistance, the chances of sustaining the yield advantages of a resistant variety improve if the management practices of farmers are fine-tuned. IPM practices, which are known to be knowledge-intensive, are often recommended to stall the development of pest resistance or secondary pests. Moreover, a longer-term goal for farmers in developing economies is greater knowledge about farming, enhanced literacy, numeracy, and mastery of skills.

Related to knowledge diffusion is a second major lacuna in the literature: impacts within communities as compared to average impacts on individual farms, as represented by survey means and estimated regression parameters. The next generation of economics studies will need to look more critically at impacts on the demand and supply of labor, health and the environment, and equity and relative poverty—which have not yet received rigorous treatment in the peer-reviewed, applied economics literature. One striking discrepancy in this literature is the general absence of gender analysis.

Thinking outside the regimen of standard statistical surveys may be worthwhile. With the assistance of leaders in local communities, it may be possible to design consensual, participatory approaches to measuring and documenting farm impacts. Carefully constructed, community-managed experiments may offer a means of superseding some types of selection bias while empowering local communities to generate their own information. Recent advances in choice experiments, such as combined revealed- and stated-preference methods, could be conducted with farmers to generate estimates of WTP for seed and seed-related information. To better understand how farmers accumulate knowledge about transgenic crops, researchers might consider analysis of formal and informal networks for disseminating seed or seed-related information in farming communities. Depending on the social structure in certain communities, community registers and diaries might be one way to supplement data collected through experimental measurement with qualitative and attitudinal information. For some types of basic monitoring data, the cost-effectiveness of using cell phones and other new technologies to collect basic data might be evaluated by research teams. Applications of more advanced valuation methods are clearly needed in order to generate more precise estimates of health and environmental impacts.

Consumers

Conclusions

Without consumer demand for biotech products, farmers will not be able to sell their crops on the market even if they adopt them. For this reason, consumer attitudes and perceptions play a fundamental role in the global debate concerning transgenic crops. Two main bodies of literature address the influence of transgenic crops on consumer behavior. The first consists of surveys designed to elicit the attitudes of consumers toward products made with transgenic crops (here we refer to these as biotech foods). Findings are generally descriptive in nature. In the second set of articles, authors exploit recent advances in stated-preference methods to estimate consumers' WTP for products that are free of transgenic ingredients. All findings in this second set are based on hypothetical, rather than observed, choices. Aside from these two groups, four studies explore the impact of biofortified crops on public health in Asia. All studies in this third group are ex ante—of use in setting priorities and gauging potential but of no use in testing hypotheses.

China is by far the most heavily studied developing economy in this literature. Authors generally concur that Chinese consumers are more accepting of biotech products than are consumers in other countries. They ascribe this result to a combination of government policy and cultural and political history that is unique to China. Thus there is a salient need for research to be conducted in other developing economies.

While there is little overt reference to the role of information in farm impact studies, a common thread across all consumer studies is the crucial role of information in preference evolution. Authors have shown that attitudes of consumers change significantly as they absorb new information, and particularly negative messages. Framing of questions is therefore of great importance, and studies will have to be periodically updated as the market changes. Relative to their counterparts in developed economies, most consumers in developing economies have serious constraints on access to information about biotech food.

Future Research Directions

Until recently few studies have sought to estimate WTP with advanced methods, and thus there is considerable room for innovation in designing future research. The most obvious limitation in this literature stems from the fact that because few biotech foods have been introduced in developing economies, consumer studies are for the most part hypothetical or ex ante. Authors have not yet applied recent approaches that combine revealed- and stated-preference approaches. Such approaches could strengthen the reliability of findings or serve for cross-validation. In common with the farm-based studies discussed previously, recent consumer studies recognize the issue of selectivity bias, or the relationship between the latent variables that predict a decision to participate (purchase) in a market and the effect of participation on its outcome.

There is generally a disconnection between farmer and consumer studies in this literature, in part because the most-studied transgenic crop is IR cotton, which is not a food crop, and few biotech foods are marketed in developing economies. Sample sizes are often considerably larger in the consumer studies compared to the farm studies, especially in industrialized economies, because it is feasible to use cheaper interview methods such as mail, phone, or Internet surveys. The effectiveness of such approaches in most developing economies is probably questionable. Even so, compared to farm surveys, the concentration of large numbers of consumers in shopping centers or markets facilitates larger samples for personal interviews.

The relevance of consumer attitudes against transgenic products and the powerful impact of media messages on consumer behavior underscore the need for more advanced analysis of consumer preferences and behavior as part of technology assessment. Analyses of consumer attitudes and behavior are crucial for gauging the potential economic impact of transgenic crops. The breadth and geographical scope of these studies has so far been quite limited, in large part because few markets for biotech foods exist.

Industry

Conclusions

Sector studies join the supply and demand sides of the market or the "industry" for the transgenic crop. In the literature dealing with industry impacts of transgenic crops in developing economies, almost all studies have been conducted ex ante using the economic surplus model. Thus estimates have largely gauged potential rather than actual benefits—providing a rough indicator of the relative magnitude and distribution of impact among actors. The economic surplus model has been combined in some cases with farm survey analysis or ex ante assessments of nutritional impacts. A new category consists of publications that examine regulatory costs and impacts on seed markets, but so far these lack a strong theoretical formulation. When considering transgenic crops, the principal modification of the economic surplus model involves the temporary monopoly conferred on the innovator through proprietary rights over the product. Thus the articles in this small group of studies are fairly homogeneous in terms of methods.

The major advantage of the economic surplus approach is that it is parsimonious with respect to data and can be used to portray the distributional effects of a range of institutional and market structures. Estimates provide a starting point for policy discussions and decisions regarding strategic investments in agricultural research.

The principal disadvantage of the economic surplus approach is that it is incomplete. The approach best depicts an industry composed of specialized, commercially oriented farmers who buy and sell in well-organized markets and grow their crop under relatively homogeneous conditions. In many developing economies this assumption is restrictive given the semisubsistence orientation and diversified production units of farmers, especially for food crops.

The quality of the underlying data is crucial for the validity of the results. In general, reliable cross-sectional, time-series data to support sector analyses of transgenic crops are not yet available in most developing economies. At present such data are probably too costly to assemble, maintain, and disseminate publicly given the information infrastructure found in most of these countries. The databases employed by researchers in China and India are exceptional in this respect.

One way that researchers have compensated for the lack of large cross-sectional, time-series databases has been to expand existing data from both primary and secondary sources using stochastic simulation. These tools assume special significance when technologies are grown by farmers in heterogeneous production environments for uncertain markets, where location and year-specific effects on productivity can generate large coefficients of variation in model parameters, including farm profits, adoption rates, and prices. If the number of input suppliers is small or markets must be segregated, risk and uncertainty in the market channel may be somewhat higher in the case of transgenic crops relative to other new crop varieties.

Future Research Directions

As demonstrated in several articles, stochastic analysis of yield, cost, and profit parameters measured on farms can be employed to better account for risk and uncertainty, serving to augment the standard economic surplus model. With respect to IR crops, sector models could be strengthened by incorporating some of the lessons learned from the application of biophysical models. For example, parameters could be specified according to estimates of damage abatement rather than productivity shifts. Biophysical models for the evolution of pest resistance can be linked to the sector model. Sector model innovations are relatively cheap if investments in good-quality farm and industry data have already been made.

The incompleteness of the basic economic surplus approach means that important pieces of information are missing from the analyses published in the first decade. Key factors that determine the likelihood that the benefits projected will in fact be earned, such as the performance and organization of input and output markets, as well as interactions among markets, have not been adequately examined. Given the significance of issues related to biosafety regulations, supply channel performance, and industrial organization in the development and diffusion of transgenic crops, quantitative analyses of these issues are especially needed. The proprietary nature of the transgenic construct has implications for input supply channels and the flow of information among economic actors (farmers, consumers, innovators, traders). To supplement sector models, and to inform the scenarios and assumptions they portray, value chain analysis and institutional analysis of industry structure should be undertaken. After all, the institutional and social context of technology introduction is often of greater significance for determining the direction and magnitude of impacts than the effectiveness of any particular trait.

Positive and negative externalities, such as the impacts of transgenic crops on the environment and health, have not yet been incorporated into these models—although it is possible to do so in theory. As has been shown with other major research questions considered in this literature, sector impact analyses have represented few crops and traits during this first decade.

Trade

Conclusions

Trade studies are of particular relevance in this topic area because the four major transgenic crops on the market (soybeans, maize, cotton, and canola) are also major, internationally traded commodities. Countries have established trade-related regulations that are more specific for biotech products than for other agricultural products. Divergence in consumer acceptance of biotech foods also means that the issue of market segregation between products made with transgenic and nontransgenic ingredients is increasingly prominent.

Applied trade studies can be divided into three sets according to the methods used by the authors. Although all three sets are based on the use of ex post international trade data, virtually all studies are ex ante analyses of the implications of adopting transgenic crops. No ex post studies have been identified in our search.

The first set of studies exploit data on bilateral trade in descriptive or accounting analyses of trade-related issues raised by the adoption of transgenic crops. Authors are concerned about the effects of adopting transgenic crops on access to global markets. Although authors may invoke conceptual frameworks, empirical applications are not grounded in an explicit model of international markets.

The other two sets represent two major categories of applied trade models. In the first category, partial equilibrium models are applied to analyze one or several sectors of the economy in a few countries, focusing on particular vertical or horizontal linkages. These studies have the advantage of flexibility, enabling the representation of a complex array of institutional and market policies. Their principal disadvantage is that they do not take into account linkages with multiple sectors or specific regulations affecting bilateral trade.

The second category employs multicountry general equilibrium models. These models provide a consistent and comprehensive structural representation of the economy and of international trade linkages. However, they are highly aggregated and based on important assumptions about the market, which limits the extent to which authors can draw policy implications.

Since 2000 trade studies have progressively improved in terms of their representation of the international economic reality. Despite the variety of approaches, studies consistently support three main findings. First, they illustrate the importance of "first-mover" advantage. Countries that do not adopt transgenic crops lose if they stay behind. Second, a number of studies highlight the risk of productivity growth in markets with inelastic demand, which benefits consumers to the detriment of adopting producers. Third, many studies demonstrate that in developing economies, potential export losses resulting from the adoption of transgenic crops are unjustified relative to the potential gains from productivity enhancement.

All three sets of studies share several limitations. First, because these are ex ante studies, they are based on assumptions that are uncertain, but the treatment of uncertainty is unsophisticated relative to some of the sector studies. Second, because the macroeconomy is the unit of analysis, they rely on a number of assumptions concerning aggregation. Some of these are problematic, such as the aggregation of nonhomogenous countries, sectors, crops, and varieties. Third, most models are based on the assumption of perfectly competitive markets, which is likely to be unrealistic given the structure of the commodity supply chains, particularly in the input sector. Fourth, there has been no explicit effort so far to reconcile results from trade studies with the reality of markets in developing countries. Market integration is not perfect in most developing nations. Many developing economies maintain policies that distort prices, so that price transmission is imperfect, the distribution of rents is asymmetric, and producers or consumers may not be directly affected by international changes.

Each set of studies has its own strengths and weaknesses. Descriptive studies can provide only limited insights with respect to quantitative conclusions, and they do not consider price effects. Some of these studies may provide useful first steps, highlighting key policy issues. Only a few published articles used partial equilibrium simulation models to evaluate the introduction of transgenic crops, and few of these focus on developing countries. Unfortunately simulations conducted in partial equilibrium frameworks rely on relatively simplistic assumptions about the adoption and productivity effects of the technology. The progressive improvement in applied general equilibrium modeling in the published literature has resulted in the deflation of the computed welfare effects associated with the introduction of the technology.

Future Research Directions

Several key methodological issues remain to be resolved in these models. Despite improvements in modeling productivity in general, other aspects—such as the roles of regional differences, labor effects, land types, and particularly seed prices—need fuller articulation. Most studies do not make an effort to adjust for the aggregation of sectors in the GTAP database. There has been no effort to model market imperfections in the input sector. Only two studies include costs of segregation for nontransgenic products. Trade-related regulations for biotech food, particularly labeling and import approval regulations, are in need of better treatment. Markets for unauthorized seeds of transgenic crops have not yet been represented, although certain countries are known to have few controls in place at their borders.

In conducting descriptive analyses of international trade, careful attention must be paid to data sources. More should be done to make sure the source of data is reliable when drawing policy conclusions. At a minimum, because of the potential for inconsistency in databases that are generally available, data points should be averaged over several years of data and the use of outliers should be avoided.

More nuanced trade models are feasible, and studies could more fully treat the regulations that affect developing countries in either partial or general equilibrium frameworks. Partial equilibrium models will be strengthened with better representation of the technology effects and improved calibration of the effects of trade-related regulations. Future studies would benefit from using improved sector models to represent the productivity effects of the technology as a basis for studying the effects of international differences in trade-related regulations. Models that incorporate properties of bilateral trade flows would improve assessments of the effects of specific regulations. In computable general equilibrium models, policies should be represented in a more realistic way. The circulation of unauthorized seed might be modeled as a type of technology spillover. Proprietary rights, combined with the structure of the seed industry, entail sector and trade models that account for imperfect competition. The potential for market segregation must also be addressed.

References

Adiyoga, W., M. Ameriana, S. R. Francisco, C. B. C. Mamaril, J. M. Yorobe Jr., and G. W. Norton. 2006. *Projected impacts of biotechnology products in Indonesia and the Philippines*. Agricultural Biotechnology Support Program II (ABSP-II) Report. Ithaca, N.Y., U.S.A.: International Service for the Acquisition of Agri-biotech Applications (ISAAA).

Aguilar, F. X., and B. Kohlmann. 2006. Willingness to consume and produce transgenic bananas in Costa Rica. *International Journal of Consumer Studies* 30 (6): 544–551.

Alston, J. M., G. W. Norton, and P. G. Pardey. 1998. *Science under scarcity: Principles and practice for agricultural research evaluation and priority setting*, 2nd ed. Wallingford, U.K.: CAB International.

Anand, A., R. C. Mittelhammer, and J. J. McCluskey. 2007. Consumer response to information and second-generation genetically modified food in India. *Journal of Agricultural and Food Industrial Organization* 5 (1): 1–18.

Anderson, K., and L. A. Jackson. 2003. Why are US and EU policies toward GMOs so different? *AgBioForum* 6 (3): 95–100.

Anderson, K., and L. A. Jackson. 2005. Some implications of GM food technology policies for Sub-Saharan Africa. *Journal of African Economies* 14 (3): 385–410.

Anderson, K., L. A. Jackson, and C. P. Nielsen. 2004. *Genetically modified rice adoption: Implications for welfare and poverty alleviation*. Working Paper 0413. Adelaide, Australia: Centre for International Economic Studies (CIES), University of Adelaide.

Anderson, K., and E. Valenzuela. 2007. The World Trade Organisation's Doha cotton initiative: A tale of two issues. *World Economy* 30 (8): 1281–1304.

Anderson, K., E. Valenzuela, and L. A. Jackson. 2008. Recent and prospective adoption of genetically modified cotton: A global computable general equilibrium analysis of economic impacts. *Economic Development and Cultural Change* 56 (2): 265–296.

Anderson, K., and S. Yao. 2003. China, GMOs and world trade in agricultural and textile products. *Pacific Economic Review* 8 (2): 157–169.

Anderson, K., C. P. Nielsen, S. Robinson, and K. Thierfelder. 2001. Estimating the global economic effects of GMOs. In The future of food: Biotechnology markets and policies in an international setting, ed. P. G. Pardey, 49–54. Washington, D.C.: International Food Policy Research Institute.

Annou, M. M., F. H. Fuller, and E. J. Wailes. 2005. Innovation dissemination and the market impacts of drought-tolerant, genetically modified rice. *International Journal of Biotechnology* 7 (1–3): 113–127.

Arunachalam, V., and S. Bala Ravi. 2003. Conceived conclusions in favour of GM cotton? A riposte to a paper in *Science. Current Science* 85 (8): 1117–1119.

Avila, A. F. D., T. R. Quirino, E. Contini, and E. L. R. Filho. 2002. Social and economic impact ex ante evaluation of Embrapa's biotechnology research products. In *Economic and social issues in agricultural biotechnology*, ed. R. E. Evenson, V. Santaniello, and D. Zilberman, 287–307. Wallingford, U.K.: CAB International.

Babu, S. C., and V. Rhoe. 2003. Assessing agricultural biotechnology: Applications of ex-ante and ex-post methods to genetically modified crops. *Asian Biotechnology and Development Review* 5 (3): 1–22.

Barwale, R. B., V. R. Gadwal, U. Zehr, and B. Zehr. 2004. Prospects for Bt cotton technology in India. *AgBioForum* 7 (1–2): 23–26.

Bennett, R., Y. Ismael, and S. Morse. 2005. Explaining contradictory evidence regarding impacts of genetically modified crops in developing countries: Varietal performance of transgenic cotton in India. *Journal of Agricultural Science* 143 (1): 35–41.

Bennett, R., S. Morse, and Y. Ismael. 2006. The economic impact of genetically modified cotton on South African smallholders: Yield, profit and health effects. *Journal of Development Studies* 42 (4): 662–677.

Bennett, R., T. Buthelezi, Y. Ismael, and S. Morse. 2003. Bt cotton, pesticides, labour and health: A case study of smallholder farmers in the Makhathini Flats, Republic of South Africa. *Outlook on Agriculture* 32 (2): 123–128.

Bennett, R., Y. Ismael, U. Kambhampati, and S. Morse. 2004a. Economic impact of genetically modified cotton in India. *AgBioForum* 7 (3): 96–100.

Bennett, R., Y. Ismael, S. Morse, and B. Shankar. 2004b. Reductions in insecticide use from adoption of Bt cotton in South Africa: Impacts on economic performance and toxic load to the environment. *Journal of Agricultural Science* 142 (6): 665–674.

Bennett, R., U. Kambhampati, S. Morse, and Y. Ismael. 2006. Farm-level economic performance of genetically modified cotton in Maharashtra, India. *Review of Agricultural Economics* 28 (1): 59–71.

Berwald, D., C. A. Carter, and G. P. Gruère. 2006. Rejecting new technology: The case of genetically modified wheat. *American Journal of Agricultural Economics* 88 (2): 432–447.

Birol, E., E. R. Villalba, and M. Smale. 2007. *Farmer preferences for Milpa diversity and genetically modified maize in Mexico: A latent class approach.* IFPRI Discussion Paper 726. Washington, D.C.: International Food Policy Research Institute.

Bouis, H. 2002. Three criteria for establishing the usefulness of biotechnology for reducing micronutrient malnutrition. *Food and Nutrition Bulletin* 23 (4): 351–353.

Brookes, G., and P. Barfoot. 2006. Global impact of biotech crops: Socio-economic and environmental effects in the first ten years of commercial use. *AgBioForum* 9 (3): 139–151.

Cabanilla, L. S. 2004. Bt corn in the Philippines: How much will farmers expect to gain? Paper presented at the PAEDA Convention, Quezon City, Philippines, Los Banos, October 12.

Cabanilla, L. S., T. Abdoulaye, and J. H. Sanders. 2005. Economic cost of non-adoption of Bt-cotton in West Africa: With special reference to Mali. *International Journal of Biotechnology* 7 (1–3): 46–61.

Carpenter, J. E., and L. P. Gianessi. 2001. *Agricultural biotechnology: Updated benefits estimates.* Washington, D.C.: National Center for Food and Agricultural Policy.

Caswell, M. F., K. O. Fuglie, and C. A. Klotz. 1994. *Agricultural biotechnology: An economic perspective.* Agricultural Economic Report 687. Washington, D.C.: Resources and Technology Division, Economic Research Service, United States Department of Agriculture.

Chern, W. S., K. Rickertsen, N. Tsuboi, and T.-T. Fu. 2002. Consumer acceptance and willingness to pay for genetically modified vegetable oil and salmon: A multiple-country assessment. *AgBioForum* 5 (3): 105–112.

Contini, E., M. J. A. Sampaio, and A. F. D. Avila. 2005. The lack of clear GMO regulation: Its impact on researchers and farmers in Brazil. *International Journal of Biotechnology* 7 (1–3): 29–45.

Crost, B., B. Shankar, R. Bennett, and S. Morse. 2007. Bias from farmer self-selection in genetically modified crop productivity estimates: Evidence from India data. *Journal of Agricultural Economics* 58 (1): 24–36.

Curtis, K. R., and K. Moeltner. 2006. Genetically modified food market participation and consumer risk perceptions: A cross-country comparison. *Canadian Journal of Agricultural Economics / Revue canadienne d'agroéconomie* 54 (2): 289–310.

Da Silveira, J. M. F., and I. d. C. Borges. 2005. An overview of the current state of agricultural biotechnology in Brazil. Paper presented at a workshop held at Bellagio, Italy, May 30–June 1.

Dawe, D., R. Robertson, and L. Unnevehr. 2002. Golden rice: What role could it play in alleviation of vitamin A deficiency? *Food Policy* 27 (5–6): 541–560.

De Groote, H., W. Overholt, J. O. Ouma, and S. Mugo. 2003. Assessing the potential impact of Bt maize in Kenya using a GIS based model. Paper presented at the International Agricultural Economics Conference, Durban, August 16–18.

Demont, M., and E. Tollens. 1999. *The economics of agricultural biotechnology: Historical and analytical framework.* Working Paper 53. Leuven, Belgium: Department of Agricultural and Environmental Economics, Katholieke Universiteit Leuven.

Deodhar, S. Y., S. Ganesh, and W. S. Chern. 2007. *Emerging markets for GM foods: An Indian perspective on consumer understanding and willingness to pay.* IIMA Working Paper 2007-06-08. Ahmedabad, India: Indian Institute of Management.

Díaz Osorio, J., R. Herrera, J. Valderrama, and J. L. Llanos Ascencio. 2004. Potential changes in the competitiveness of maize growers in Central Chile through the use of transgenic seed (Bt and RR). *Spanish Journal of Agricultural Research* 2 (2): 145–156.

Edmeades, S., and M. Smale. 2006. A trait-based model of the potential demand for a genetically engineered food crop in a developing economy. *Agricultural Economics* 35 (3): 351–361.

Elbehri, A., and S. Macdonald. 2004. Estimating the impact of transgenic *Bt* cotton on West and Central Africa: A general equilibrium approach. *World Development* 32 (12): 2049–2064.

Falck-Zepeda, J., J. Cohen, and J. Komen. 2003. Impact assessment and agricultural biotechnology-research methodologies for developing, emerging and transition economies. In *Accessing agricultural biotechnology in emerging economies.* Paris: Organisation for Economic Cooperation and Development.

Falck-Zepeda, J., D. Horna, and M. Smale. 2007. *The economic impact and the distribution of benefits and risk from the adoption of insect resistant (Bt) cotton in West Africa.* IFPRI Discussion Paper 718. Washington, D.C.: International Food Policy Research Institute.

Falck-Zepeda, J., G. Traxler, and R. G. Nelson. 1999. *Rent creation and distribution from the first three years of planting Bt cotton.* ISAAA Briefs 14. Ithaca, N.Y., U.S.A.: International Service for the Acquisition of Agri-biotech Applications (ISAAA).

———. 2000a. Rent creation and distribution from biotechnology innovations: The case of Bt cotton and herbicide-tolerant soybeans in 1997. *Agribusiness* 16: 21–32.

———. 2000b. Surplus distribution from the introduction of a biotechnology innovation. *American Journal of Agricultural Economics* 82: 360–369.

Falck-Zepeda, J., N. Barreto-Triana, I. Baquero-Haeberlin, E. Espitia-Malagón, H. Fierro-Guzmán, and N. López. 2006. *An exploration of the potential benefits of integrated pest management systems and the use of insect resistant potatoes to control the Guatemalan tuber moth (*Tecia solanivora *Povolny) in Ventaquemada, Colombia.* Environment and Production Technology Division Discussion Paper 152. Washington, D.C.: International Food Policy Research Institute.

Fan, C., J. Li, R. Hu, and C. Zhang. 2002. Effects of planting of transgenic Bt (*Bacillus thuringiensis*) insect-resistant cotton on herbicide application. *China Rural Survey* (5): 2–10.

Felloni, F., J. Gilbert, T. I. Wahl, and P. Wandschneider. 2003. Trade policy, biotechnology and grain self-sufficiency in China. *Agricultural Economics* 28 (3): 173–186.

Fernandez-Cornejo, J., and M. F. Caswell. 2006. *The first decade of genetically engineered crops in the United States.* Economic Information Bulletin 11. Washington, D.C.: Economic Research Service, United States Department of Agriculture.

Fok, M., W. Liang, G. Wang, and Y. Wu. 2005. Diffusion du coton génétiquement modifié en Chine: Leçons sur les facteurs et limites d'un succès. *Economie Rurale* 285 (January–February): 5–32.

Food and Agriculture Organization of the United Nations (FAO). 2004. *The state of food and agriculture 2003 2004: Agricultural biotechnology meeting the needs of the poor?* Rome.

Fransen, L. 2006. The debate distilled: A review of the literature on social and economic impacts of transgenic crops. IFPRI consultant report. Washington, D.C.: International Food Policy Research Institute.

Frisvold, G. B., J. M. Reeves, and R. Tronstad. 2006. Bt cotton adoption in the United States and China: International trade and welfare effects. *AgBioForum* 9 (2): 69–78.

Frohlich, X. 2005. Background paper to the Oxfam prospectus: Transgenic cotton project. Summer internship report. Washington, D.C.: Oxfam.

Gouse, M., J. Kirsten, and L. Jenkins. 2003. Bt cotton in South Africa: Adoption and the impact on farm incomes amongst small-scale and large scale farmers. *Agrekon* 42 (1): 15–28.

Gouse, M., C. E. Pray, and D. Schimmelpfennig. 2004. The distribution of benefits from Bt cotton adoption in South Africa. *AgBioForum* 7 (4): 187–194.

Gouse, M., J. Kirsten, B. Shankar, and C. Thirtle. 2005a. Bt cotton in KwaZulu Natal: Technological triumph but institutional failure. *AgBiotechNet* 7 (134): 1–7.

Gouse, M., C. E. Pray, J. Kirsten, and D. Schimmelpfennig. 2005b. A GM subsistence crop in Africa: The case of Bt white maize in South Africa. *International Journal of Biotechnology* 7 (1–3): 84–94.

Gouse, M., C. E. Pray, D. Schimmelpfennig, and J. Kirsten. 2006. Three seasons of subsistence insect-resistant maize in South Africa: Have smallholders benefited? *AgBioForum* 9 (1): 15–22.

Govindasamy, R., B. Onyango, W. Hallman, H.-M. Jang, and V. S. Puduri. 2004. *Public approval of plant and animal biotechnology in Korea: An ordered probit analysis.* Research Report WP1104-016. New Brunswick, N.J.: Food Policy Institute, Rutgers, the State University of New Jersey.

Griffin, K. 1974. *The political economy of agrarian change: An essay on the Green Revolution.* Cambridge, Mass., U.S.A.: Harvard University Press.

Gruère, G. P., A. Bouët, and S. Mevel. 2007. *Genetically modified food and international trade: The case of India, Bangladesh, Indonesia and the Philippines.* IFPRI Discussion Paper 740. Washington, D.C.: International Food Policy Research Institute.

Hareau, G. G., B. F. Mills, and G. W. Norton. 2006. The potential benefits of herbicide-resistant transgenic rice in Uruguay: Lessons for small developing countries. *Food Policy* 31 (2): 162–179.

Hareau, G. G., B. F. Mills, G. W. Norton, and D. Bosch. 2002. The economic impact of genetically modified organisms in small developing countries. Paper prepared for the annual meeting of the American Agricultural Economics Association, Long Beach, Calif., U.S.A., July 28–31.

Hareau, G. G., G. W. Norton, B. F. Mills, and E. Peterson. 2005. Potential benefits of transgenic rice in Asia: A general equilibrium analysis. *Quarterly Journal of International Agriculture* 44 (3): 229–246.

Hayami Y., and R. W. Herdt. 1977. Market price effects of technological change on income distribution in semisubsistence agriculture. *American Journal of Agricultural Economics* 59 (2): 245–256.

Hazell, P., and C. Ramaswamy. 1991. *The Green Revolution reconsidered: The impact of high-yielding rice varieties in South India.* Baltimore, Md., U.S.A.: Johns Hopkins University Press.

Hertel, T. W., ed. 1997. *Global trade analysis: Modeling and applications.* Cambridge: Cambridge University Press.

Ho, P., and E. B. Vermeer. 2004. Food safety concerns and biotechnology: Consumers' attitudes to genetically modified products in urban China. *AgBioForum* 7 (4): 158–175.

Ho, P., E. B. Vermeer, and J. H. Zhao. 2006. Biotechnology and food safety in China: Consumers' acceptance or resistance? *Development and Change* 37 (1): 227–253.

Hofs, J.-L., M. Fok, and M. Vaissayre. 2006. Impact of Bt cotton adoption on pesticide use by smallholders: A 2-year survey in Makhatini Flats (South Africa). *Crop Protection* 25 (9): 984–988.

Hu, W. Y. 2006. Use of spike models in measuring consumers' willingness to pay for non-GM oil. *Journal of Agricultural and Applied Economics* 38 (3): 525–538.

Hu, W. Y., and K. Chen. 2004. Can Chinese consumers be persuaded? The case of genetically modified vegetable oil. *AgBioForum* 7 (3): 124–132.

Hu, W. Y., F. Zhong, and Y. Ding. 2006. Actual media reports on GM foods and Chinese consumers' willingness to pay for GM soybean oil. *Journal of Agricultural and Resource Economics* 31 (2): 376–390.

Huang, J., R. Hu, C. Fan, C. E. Pray, and S. Rozelle. 2002a. Bt cotton benefits, costs, and impacts in China. *AgBioForum* 5 (4): 153–166.

Huang, J., R. Hu, S. Rozelle, F. Qiao, and C. Pray. 2002b. Transgenic varieties and productivity of smallholder cotton farmers in China. *Australian Journal of Agricultural and Resource Economics* 46 (3): 367–387.

Huang, J., R. Hu, Q. Wang, J. Keeley, and J. Falck-Zepeda. 2002c. Agricultural biotechnology development, policy and impact in China. *Economic and Political Weekly* 37 (27): 2756–2761.

Huang, J., R. Hu, C. E. Pray, F. Qiao, and S. Rozelle. 2003. Biotechnology as an alternative to chemical pesticides: A case study of Bt cotton in China. *Agricultural Economics* 29 (1): 55–67.

Huang, J., R. Hu, H. van Meijl, and F. van Tongeren. 2004. Biotechnology boosts to crop productivity in China: Trade and welfare implications. *Journal of Development Economics* 75 (1): 27–54.

Huang, J., R. Hu, S. Rozelle, and C. E. Pray. 2005. Insect-resistant GM rice in farmers' fields: Assessing productivity and health effects in China. *Science* 308 (5722): 688–690.

Huang, J., H. Qiu, J. Bai, and C. E. Pray. 2006. Awareness, acceptance of and willingness to buy genetically modified foods in urban China. *Appetite* 46 (2): 144–151.

Huesing, J., and L. English. 2004. The impact of Bt crops on the developing world. *AgBioForum* 7 (1–2): 84–95.

Ismael, Y., R. Bennett, and S. Morse. 2001. Farm level impact of Bt cotton in South Africa. *Biotechnology and Development Monitor* 48 (December): 15–19.

———. 2002a. Benefits from Bt cotton use by smallholder farmers in South Africa. *AgBioForum* 5 (1): 1–5.

———. 2002b. Farm-level economic impact of biotechnology: Smallholder Bt cotton farmers in South Africa. *Outlook on Agriculture* 31 (2): 107–111.

Ismael, Y., L. Beyers, C. Thirtle, and J. Piesse. 2002. Efficiency effects of Bt cotton adoption by smallholders in Makhathini Flats, KwaZulu-Natal, South Africa. In *Economic and social issues in agricultural biotechnology,* ed. R. E. Evenson, V. Santaniello, and D. Zilberman, 325–349. Wallingford, U.K.: CAB International.

Jan, M., T.-T. Fu, and C. L. Huang. 2007. A conjoint/logit analysis of consumers' responses to genetically modified tofu in Taiwan. *Journal of Agricultural Economics* 58 (2): 330–347.

Kambhampati, U., S. Morse, R. Bennett, and Y. Ismael. 2005. Perceptions of the impacts of genetically modified cotton varieties: A case study of the cotton industry in Gujarat, India. *AgBioForum* 8 (2–3): 161–171.

Kirsten, J., and M. Gouse. 2003. The adoption and impact of agricultural biotechnology in South Africa. In *The economic and environmental impacts of agbiotech: A global perspective,* ed. N. Kalaitzandonakes, 243–260. New York: Kluwer Academic/Plenum.

Klotz-Ingram, C., S. Jans, and J. Fernandez-Cornejo. 1999. Farm-level production effects related to the adoption of genetically modified cotton for pest management. *AgBioForum* 2 (2): 73–84.

Kolady, D. E., and W. Lesser. 2006. Who adopts what kind of technologies? The case of Bt eggplant in India. *AgBioForum* 9 (2): 94–103.

———. 2007. Can owners afford humanitarian donations in agbiotech—The case of genetically engineered eggplant in India. *Electronic Journal of Biotechnology* 11 (2): 1–8.

———. 2008. Is genetically engineered technology a good alternative to pesticide use: The case of GE eggplant in India. *International Journal of Biotechnology* 10 (2–3): 132–147.

Kontoleon, A. 2003. Accounting for consumer heterogeneity in preferences over GM foods: An application of the latent market segmentation model. Paper presented at the Fourth BIOECON Workshop on the Economics of Biodiversity Conservation, Venice, Italy, August 28–29.

Krishna, V. V., and M. Qaim. 2007. Estimating the adoption of Bt eggplant in India: Who benefits from public-private partnership? *Food Policy* 32 (5–6): 523–543.

Kuosmanen, T., D. Pemsl, and J. Wesseler. 2006. Specification and estimation of production functions involving damage control inputs: A two-stage, semiparametric approach. *American Journal of Agricultural Economics* 88 (2): 499–511.

Langyintuo, A. S., and J. Lowenberg-DeBoer. 2006. Potential regional trade implications of adopting Bt cowpea in West and Central Africa. *AgBioForum* 9 (2): 111–120.

Li, Q., K. R. Curtis, J. J. McCluskey, and T. I. Wahl. 2002. Consumer attitudes toward genetically modified foods in Beijing, China. *AgBioForum* 5 (4): 145–152.

Lin, W., A. Somwaru, F. Tuan, J. Huang, and J. Bai. 2006. Consumers' willingness to pay for biotech foods in China: A contingent valuation approach. *AgBioForum* 9 (3): 166–179.

Lipton, R., and R. Longhurst. 1989. *New seeds and poor people.* Baltimore, Md., U.S.A.: Johns Hopkins University Press.

Lusk, J. L., M. Jamal, L. Kurlander, M. Roucan, and L. Taulman. 2005. A meta analysis of genetically modified food valuation studies. *Journal of Agricultural and Resource Economics* 30 (1): 28–44.

Mamaril, C. B. C., and G. W. Norton. 2006. Economic evaluation of transgenic pest resistant rice in the Philippines and Vietnam. *Quarterly Journal of International Agriculture* 45 (2): 127–144.

Manalo, A. J., and G. P. Ramon. 2007. The cost of product development of Bt corn event MON810 in the Philippines. *AgBioForum* 10 (1): 19–32.

Marra, M. C., P. G. Pardey, and J. M. Alston. 2002. The payoffs to transgenic field crops: An assessment of the evidence. *AgBioForum* 5 (2): 43–50.

Massieu, Y., L. R. González, M. Chauvet, Y. Castañeda, and R. E. Barajas. 2000. Transgenic potatoes for small-scale farmers: A case study in Mexico. *Biotechnology and Development Monitor* 41 (March): 6–10.

McCluskey, J. J., K. M. Grimsrud, and T. I. Wahl. 2006. Comparison of consumer responses to genetically modified foods in Asia, North America, and Europe. In *Regulating agricultural biotechnology: Economics and policy,* ed. R. E. Just, J. M. Alston, and D. Zilberman, 227–240. Natural Resource Management and Policy. New York: Springer-Verlag.

Mellor, J. W., and B. F. Johnston. 1984. The world food equation: Interrelations among development, employment, and food consumption. *Journal of Economic Literature* 22 (2): 524–31.

Morse, S., R. M. Bennett, and Y. Ismael. 2005a. Bt-cotton boosts the gross margin of small-scale cotton producers in South Africa. *International Journal of Biotechnology* 7 (1–3): 72–83.

———. 2005b. Comparing the performance of official and unofficial genetically modified cotton in India. *AgBioForum* 8 (1): 1–6.

———. 2005c. Genetically modified insect resistance in cotton: Some farm level economic impacts in India. *Crop Protection* 24 (5): 433–440.

———. 2006. Environmental impact of genetically modified cotton in South Africa. *Agriculture Ecosystems and Environment* 117 (4): 277–289.

———. 2007. Inequality and GM crops: A case study of Bt cotton in India. *AgBioForum* 10 (1): 44–50.

Moschini, G. 2001. Biotech—Who wins? Economic benefits and costs of biotechnology innovations in agriculture. *Estey Centre Journal of International Law and Trade Policy* 2 (1): 93–117.

Moschini, G., and H. Lapan. 1997. Intellectual property rights and the welfare effects of agricultural R&D. *American Journal of Agricultural Economics* 79 (4): 1229–1242.

Moschini, G., H. Lapan, and A. Sobolevsky. 2000. Roundup Ready(R) soybeans and welfare effects in the soybean complex. *Agribusiness* 16 (1): 33–55.

Mucci, A., G. Hough, and C. Ziliani. 2004. Factors that influence purchase intent and perceptions of genetically modified foods among Argentine consumers. *Food Quality and Preference* 15 (6): 559–567.

Muñoz, R. 2004. The expansion of biotechnologically modified soybean in Argentina: Economical aspects and considerations. In *Proceedings VII World Soybean Research Conference, IV International Soybean Processing and Utilization Conference, III Congresso Brasileiro de Soja (Brazilian Soybean Congress), Foz do Iguassu, PR, Brazil, 29 February–5 March, 2004,* ed. F. Moscardi, C. B. Hoffmann-Campo, O. F. Saraiva, P. R. Galerani, F. C. Krzyzanowski, and M. C. Carrao-Panizzi, 551–558. Foz do Iguassu, Brazil: Brazilian Agricultural Research Corporation, National Soybean Research Center.

Murugkar, M., B. Ramaswami, and M. Shelar. 2007. Competition and monopoly in the Indian cotton seed market. *Economic and Political Weekly* 42 (37): 3781–3789.

Naik, G., M. Qaim, A. Subramanian, and D. Zilberman. 2005. Bt cotton controversy: Some paradoxes explained. *Economic and Political Weekly* 40 (15): 1514–1517.

Narayanamoorthy, A., and S. S. Kalamkar. 2006. Is Bt cotton cultivation economically viable for Indian farmers? An empirical analysis. *Economic and Political Weekly* 41 (26): 2716–2724.

Nielsen, C. P., and K. Anderson. 2001a. Global market effects of alternative European responses to genetically modified organisms. *Review of World Economics* 137 (2): 320–346.

———. 2001b. GMOs, trade policy, and welfare in rich and poor countries. In *Quantifying the impact of technical barriers to trade,* ed. J. Maskus and J. Wilson, 155–184. Ann Arbor: University of Michigan Press.

Nielsen, C. P., S. Robinson, and K. Thierfelder. 2001. Genetic engineering and trade: Panacea or dilemma for developing countries. *World Development* 29 (8): 1307–1324.

Nielsen, C. P., K. Thierfelder, and S. Robinson. 2003. Consumer preferences and trade in genetically modified foods. *Journal of Policy Modeling* 25 (8): 777–794.

Nuffield Council on Bioethics. 2004. *The use of genetically modified crops in developing countries: A follow-up discussion paper.* London.

Onyango, B., R. Govindasamy, W. Hallman, H.-M. Jang, and V. S. Puduri. 2006. Consumer acceptance of genetically modified foods in South Korea: Factor and cluster analysis. *Journal of Agribusiness* 24 (1): 61–78.

Organisation for Economic Co-operation and Development (OECD). 2000. *Modern biotechnology and agricultural markets: A discussion of selected issues.* Research Report. Paris: OECD Committee for Agriculture.

Paarlberg, R. 2006. Are genetically modified (GM) crops a commercial risk for Africa? *International Journal of Technology and Globalisation* 2 (1–2): 81–92.

Pachico, D., and M. M. Wolf. 2004. Attitudes towards GM food in Colombia. In *Consumer acceptance of genetically modified foods,* ed. R. E. Evenson and V. Santaniello, 155–161. Wallingford, U.K.: CAB International.

Pemsl, D., H. Waibel, and A. P. Gutierrez. 2005. Why do some Bt-cotton farmers in China continue to use high levels of pesticides? *International Journal of Agricultural Sustainability* 3 (1): 44–56.

Pemsl, D., H. Waibel, and J. Orphal. 2004. A methodology to assess the profitability of Bt-cotton: Case study results from the state of Karnataka, India. *Crop Protection* 23 (12): 1249–1257.

Penna, J. A., and D. Lema. 2003. The adoption of herbicide tolerant soybeans in Argentina: An economic analysis. In *The economic and environmental impacts of agbiotech: A global perspective,* ed. N. Kalaitzandonakes, 203–221. New York: Kluwer Academic/Plenum.

Pinstrup-Andersen, P. 1979. The market price effect and the distribution of economic benefits from new technology. *European Review of Agricultural Economics* 6 (1): 17–46.

Plastina, A., and K. Giannakas. 2007. Market and welfare effects of GMO introduction in small open economies. *AgBioForum* 10 (2): 104–123.

Pray, C. E., B. Ramaswami, and P. Bengali. 2005. The cost of biosafety regulations: The Indian experience. *Quarterly Journal of International Agriculture* 44 (3): 267–289.

Pray, C. E., D. Ma, J. Huang, and F. Qiao. 2001. Impact of Bt cotton in China. *World Development* 29 (5): 813–825.

Pray, C. E., B. Ramaswami, J. Huang, R. Hu, P. Bengali, and H. Zhang. 2006. Costs and enforcement of biosafety regulations in India and China. *International Journal of Technology and Globalisation* 2 (1–2): 137–157.

Price, G. K., W. Lin, J. Falck-Zepeda, and J. Fernandez-Cornejo. 2003. *Size and distribution of market benefits from adopting biotech crops.* Technical Bulletin 1906. Washington, D.C.: Economic Research Service, United States Department of Agriculture.

Purcell, J. P., and F. J. Perlak. 2004. Global impact of insect-resistant (Bt) cotton. *AgBioForum* 7 (1–2): 27–30.

Qaim, M. 1999. Potential benefits of agricultural biotechnology: An example from the Mexican potato sector. *Review of Agricultural Economics* 21 (2): 390–408.

———. 2001. A prospective evaluation of biotechnology in semi-subsistence agriculture. *Agricultural Economics* 25 (2–3): 165–175.

———. 2003. Bt cotton in India: Field trial results and economic projections. *World Development* 31 (12): 2115–2127.

Qaim, M., and A. de Janvry. 2003. Genetically modified crops, corporate pricing strategies, and farmers' adoption: The case of Bt cotton in Argentina. *American Journal of Agricultural Economics* 85 (4): 814–828.

Qaim, M., and A. de Janvry. 2005. Bt cotton and pesticide use in Argentina: Economic and environmental effects. *Environmental and Development Economics* 10 (2): 179–200.

Qaim, M., and G. Traxler. 2005. Roundup Ready soybeans in Argentina: Farm level and aggregate welfare effects. *Agricultural Economics* 32 (1): 73–86.

Qaim, M., and J. von Braun. 1998. *Crop biotechnology in developing countries: A conceptual framework for ex ante economic analyses.* ZEF Discussion Papers on Development Policy 3. Bonn, Germany: Center for Development Research.

Qaim, M., and D. Zilberman. 2003. Yield effects of genetically modified crops in developing countries. *Science* 299 (5608): 900–902.

Qaim, M., E. J. Cap, and A. de Janvry. 2003. Agronomics and sustainability of transgenic cotton in Argentina. *AgBioForum* 6 (1–2): 41–47.

Qaim, M., A. Subramanian, G. Naik, and D. Zilberman. 2006. Adoption of Bt cotton and impact variability: Insights from India. *Review of Agricultural Economics* 28 (1): 48–58.

Qayum, A., and K. Sakkhari. 2006. *Bt cotton in Andhra Pradesh—A three year assessment: The first sustained independent scientific study of Bt cotton in India.* Andhra Pradesh, India: Deccan Development Society.

Raney, T. 2006. Economic impact of transgenic crops in developing countries. *Current Opinion in Biotechnology* 17 (2): 174–178.

Sahai, S., and S. Rehman. 2003. Performance of Bt cotton: Data from first commercial crop. *Economic and Political Weekly* 38 (30): 3139–3141.

———. 2004. Bt-cotton, 2003–2004: Fields swamped with illegal variants. *Economic and Political Weekly* 39 (26): 2673–2674.

Scatasta, S., J. Wesseler, and M. Demont. 2006. *A critical assessment of methods for analysis of social welfare impacts of genetically modified crops: A literature survey.* Mansholt Working Paper MWP-27. Wageningen, The Netherlands: Mansholt Graduate School of Social Sciences.

Schaper, M., A. M. Hernández, M. A. Oliver, and M. A. Vélez. 2001. *Cinco estudios suramericanos sobre comercio y ambiente.* Quito, Ecuador: Fundación Futuro Latinoamericano.

Shankar, B., and C. Thirtle. 2005. Pesticide productivity and transgenic cotton technology: The South African smallholder case. *Journal of Agricultural Economics* 56 (1): 97–116.

Shoemaker, R., J. Harwood, K. Day-Rubenstein, T. Dunahay, P. Heisey, L. Hoffman, C. Klotz-Ingram, W. Lin, L. Mitchell, W. McBride, and J. Fernandez-Cornejo. 2001. *Economic issues in agricultural biotechnology.* Agriculture Information Bulletin 762. Washington, D.C.: Resource Economic Division, Economic Research Service, United States Department of Agriculture.

Smyth, S., W. A. Kerr, and K. A. Davey. 2006. Closing markets to biotechnology: Does it pose an economic risk if markets are globalised? *International Journal of Technology and Globalisation* 2 (3–4): 377–389.

Sobolevsky, A., G. Moschini, and H. Lapan. 2005. Genetically modified crops and product differentiation: Trade and welfare effects in the soybean complex. *American Journal of Agricultural Economics* 87 (3): 621–644.

Stein, A. J., P. Nestel, J. V. Meenakshi, M. Qaim, H. P. S. Sachdev, and Z. A. Bhutta. 2007. Plant breeding to control zinc deficiency in India: How cost-effective is biofortification? *Public Health Nutrition* 10 (5): 492–501.

Stone, S., A. Matysek, and A. Dooling. 2002. *Modelling possible impacts of GM crops on Australian trade.* Staff Research Paper. Melbourne, Australia: Productivity Commission.

Thirtle, C., L. Beyers, Y. Ismael, and J. Piesse. 2003. Can GM-technologies help the poor? The impact of Bt cotton in Makhathini Flats, KwaZulu-Natal. *World Development* 31 (4): 717–732.

Toenniessen, G. H., J. C. O'Toole, and J. DeVries. 2003. Advances in plant biotechnology and its adoption in developing countries. *Current Opinion in Plant Biology* 6 (2): 191–198.

Traxler, G., and S. Godoy-Avila. 2004. Transgenic cotton in Mexico. *AgBioForum* 7 (1–2): 57–62.

Traxler, G., S. Godoy-Avila, J. Falck-Zepeda, and J. J. Espinoza-Arellano. 2003. Transgenic cotton in Mexico: Economic and environmental impacts of the first generation biotechnologies. In *The economic and environmental impacts of agbiotech: A global perspective,* ed. N. Kalaitzandonakes, 183–202. New York: Kluwer Academic/Plenum.

Trigo, E. J., and E. J. Cap. 2006. *Ten years of genetically modified crops in Argentine agriculture.* Buenos Aires: Argentine Council for Information and Development of Biotechnology (ArgenBio).

Trigo, E. J., D. Chudnovsky, E. J. Cap, and A. López. 2002a. *Genetically modified crops in Argentine agriculture: An open ended story.* Buenos Aires: Libros del Zorzal.

Trigo, E. J., G. Traxler, C. E. Pray, and R. G. Echeverría. 2002b. *Agricultural biotechnology and rural development in Latin America and the Caribbean: Implications for IDB lending.* Washington, D.C.: Inter-American Development Bank.

Tripp, R. 2001. Can biotechnology reach the poor? The adequacy of information and seed delivery. *Food Policy* 26 (3): 249–264.

Van Meijl, H., and F. van Tongeren. 2004. International diffusion of gains from biotechnology and the European Union's Common Agricultural Policy. *Agricultural Economics* 31: 307–316.

Vitale, J., T. Boyer, R. Uaiene, and J. H. Sanders. 2007. The economic impacts of introducing Bt technology in smallholder cotton production systems of West Africa: A case study from Mali. *AgBioForum* 10 (2): 71–84.

Wu, F. 2006. Mycotoxin reduction in Bt corn: Potential economic, health and regulatory impacts. *Transgenic Research* 15 (3): 277–289.

Wu, F., and W. Butz. 2004. *The future of genetically modified crops: Lessons from the Green Revolution.* Santa Monica, Calif., U.S.A.: RAND Corporation.

Xu, J., Z. You, W. Wang, and Y. Yang. 2004. Economic analysis of Bt cotton planting in Jiangsu. *Journal of Yangzhou University, Agricultural and Life Sciences Edition* 25 (3): 65–69.

Yang, P. Y., M. Iles, S. Yan, and F. Jolliffe. 2005. Farmers' knowledge, perceptions and practices in transgenic Bt cotton in small producer systems in Northern China. *Crop Protection* 24 (3): 229–239.

Yorobe, J. M., Jr., and C. B. Quicoy. 2006. Economic impact of Bt corn in the Philippines. *Philippine Agricultural Scientist* 89 (3): 258–267.

Zhang, X. 2005. Chinese consumers' concerns about food safety: Case of Tianjin. *Journal of International Food and Agribusiness Marketing* 17 (1): 57–69.

Zhong, F., M. A. Marchant, Y. Ding, and K. Lu. 2002. GM foods: A Nanjing case study of Chinese consumers' awareness and potential attitudes. *AgBioForum* 5 (4): 136–144.

Zhou, F., and W. Tian. 2003. Consumer perceptions and attitudes toward GM food and their determinants: Case of data from Beijing. *China Agricultural Economic Review* 1 (3): 266–293.

Zimmermann, R., and M. Qaim. 2004. Potential health benefits of Golden Rice: A Philippine case study. *Food Policy* 29 (2): 147–168.

About the Authors

José Falck-Zepeda is a research fellow in the Environment and Production Technology Division of the International Food Policy Research Institute (IFPRI). He is also an affiliate professor at the University of the Philippines–Los Baños and the Escuela Agrícola Panamericana (Zamorano University), as well as a leader of the policy team at the Program for Biosafety Systems. He has a Ph.D. in agricultural economics from Auburn University. His areas of specialization include production economics and policy analysis, socioeconomic assessment of agricultural biotechnologies and other technologies, economics of biosafety regulations, and assessments of R&D capacity in developing countries.

Guillaume Gruère is a research fellow in IFPRI's Environment and Production Technology Division. He has a Ph.D. in agricultural and resource economics from the University of California, Davis. His work focuses mostly on marketing and trade-related regulations of genetically modified (GM) products and their effects on biosafety and biotechnology policymaking in developing countries. One of his articles on GM food marketing and trade policies received an award from the American Agricultural Economics Association (AAEA) in 2004.

Daniela Horna is a postdoctoral fellow in IFPRI's Environment and Production Technology Division. She has a Ph.D. in agricultural economics and social sciences from the University of Hohenheim, Germany. She has gained international experience working on the evaluation of production technologies and their contribution to household welfare in Latin America and West and East Africa. Her current research work focuses on the economics of agricultural biodiversity and in situ and ex situ conservation of plant genetic resources.

Hannah Jones was a research assistant in IFPRI's Environment and Production Technology Division from 2007 to 2008. She has a bachelor's degree in international relations from Tufts University. At IFPRI, she managed the farm-level and market databases on India, Kenya, and Mali and helped in the economic analysis of the seed systems projects in these countries. She also trained collaborators in Mali in data management and processing using STATA and compiled a wider range of literature related to biodiversity and biotechnology studies for the development of a bibliographic database.

Ira Matuschke is an agricultural economist and has a Ph.D. from the University of Hohenheim, Germany. In the past, she worked as a consultant for the International Food Policy Research Institute and the International Fund for Agricultural Development. Her areas of interest are the adoption of modern agricultural technologies and social networks in rural innovation systems.

Latha Nagarajan is a research associate in the department of food, resources, and agricultural economics at Rutgers, the State University of New Jersey. She has a Ph.D. in agricultural economics from the University of Minnesota. Prior to joining Rutgers, she was a postdoctoral fellow in IFPRI's Environment and Production Technology Division. While at IFPRI, she was responsible for implementing projects on seed systems in India and Kenya. At Rutgers, she studies the economic impact of developing and spreading drought-tolerant rice varieties in Asian countries as well as the impact of private-sector investment in agricultural R&D in South Asia and Sub-Saharan Africa.

Melinda Smale is a researcher in agriculture and trade at Oxfam America. She has a Ph.D. in agricultural economics from the University of Maryland, College Park. From 2002 to 2008, she was a senior research fellow in IFPRI's Environment and Production Technology Division. While at IFPRI, she led a global research program—initially a joint program with Bioversity International in Rome, Italy—aimed at promoting the sustainable use of crop genetic resources in developing agriculture. The program addressed the impacts of biotech crops, agricultural biodiversity, local seed markets, and underutilized crops. She is an honorary fellow with Bioversity International and serves on the advisory committee of the Collaborative Crops Research Program of the McKnight Foundation and on the editorial committees of several journals.

Indira Yerramareddy is an information and knowledge specialist in IFPRI's Communications Division. She has a master's degree in food science and nutrition from

Sri Venkateswara University, Tirupati, India. Her work at IFPRI focuses mostly on conducting database analysis, data collection and maintenance, online literature reviews, preparation of datasets for the web, and programming activities contributing to the dissemination of IFPRI research and expanding the reach and use of such research. She has developed web-based bibliographies on economic methods related to crop biotech and genetic resources.

Patricia Zambrano is a research analyst in IFPRI's Environment and Production Technology Division. She has a master's degree in economics from the University of California, Davis. Her work focuses on the evaluation of pro-poor crop biotechnologies and the economics of biosafety regulation programs. As part of this research agenda, she led a research project on the social and economic impacts of Bt cotton in small farm households in Colombia. She also led the development of web-based bibliographies on economic methods related to crop biotech and genetic resources.

Made in the USA
Monee, IL
28 February 2026

45072218R00075